PRINCIPLES OF
SPACE-TIME-MATTER

Cosmology, Particles and Waves in
Five Dimensions

PRINCIPLES OF
SPACE-TIME-MATTER

Cosmology, Particles and Waves in
Five Dimensions

Paul S Wesson
James M Overduin

Towson University, USA & Johns Hopkins University, USA

W**e** **World Scientific**

NEW JERSEY · LONDON · SINGAPORE · BEIJING · SHANGHAI · HONG KONG · TAIPEI · CHENNAI · TOKYO

Published by

World Scientific Publishing Co. Pte. Ltd.

5 Toh Tuck Link, Singapore 596224

USA office: 27 Warren Street, Suite 401-402, Hackensack, NJ 07601

UK office: 57 Shelton Street, Covent Garden, London WC2H 9HE

Library of Congress Cataloging-in-Publication Data

Names: Wesson, Paul S., author. | Overduin, J. M. (James Martin), 1965–   author.

Title: Principles of space-time-matter : cosmology, particles and waves in five dimensions /
    Paul S. Wesson, James M. Overduin, Towson University, USA & Johns Hopkins University.

Description: Singapore ; Hackensack, NJ : World Scientific Publishing Co. Pte. Ltd., [2019] |
    Includes bibliographical references and index.

Identifiers: LCCN 2018051422| ISBN 9789813235779 (alk. paper) | ISBN 9813235772 (alk. paper)

Subjects: LCSH: Relativity (Physics) | Space and time. | Particles (Nuclear physics) |
    Wave mechanics.

Classification: LCC QC173.59.S65 W4485 2019 | DDC 530.11--dc23

LC record available at https://lccn.loc.gov/2018051422

British Library Cataloguing-in-Publication Data

A catalogue record for this book is available from the British Library.

For any available supplementary material, please visit
https://www.worldscientific.com/worldscibooks/10.1142/10871#t=suppl

Preface

by First Author

The theory of Space-Time-Matter uses the geometry of the fifth dimension to explain the matter in the world. It has something in common with Einstein's 4D theory of general relativity and its 5D extensions due to Kaluza and Klein. However, STM has an energy-momentum tensor which is derived from an extra dimension that is not rolled up or compactified to an unobservably small size. In fact, we see evidence of the fifth dimension in things from the mass of an elementary particle to the density of the cosmological fluid. In this way, STM theory fulfills Einstein's theorem of transmuting the "base wood" of matter to the "fine marble" of algebra.

General relativity is an excellent theory, but it does not explain the origin of the stuff with which it deals, namely matter. Einstein realized this, though originally his "dream" was unknown to my colleagues and I. However, it was apparent that the physical quantities like mass and density could be turned into lengths and geometrized using the fundamental contexts; and that the 4D equations could be turned into expressions which combined the fields with their material sources given the appropriate extra algebra. Hence the fifth dimension. In 1992, we calculated the properties of matter using the fifth dimension for three physical systems, including the standard cosmologies, and found agreement with the conventional results. Soon after, the canonical metric was introduced, which simplified the calculations and provided an embedding in apparently empty 5D space for all solutions of the 4D Einstein equations with a finite cosmological constant or vacuum density. The initial group of workers involved in those calculations included (in alphabetical order) H. Liu, B. Mashhoon, J.M. Overduin, and J. Ponce de Leon. This nucleus grew, and eventually numbered about 20 people. It is slightly ironic that we were in full swing with our physical calculations before it was pointed out to us that there is actually a mathematical result which guarantees their validity in principle. This is Campbell's embedding theorem, which was sketched by him in a book of 1926, but lay largely forgotten until it was realized that it implies the recovery of Einstein's 4D equations with matter from the 5D Ricci equations in vacuum. The full 5D field equations describe gravitational, electromagnetic and scalar fields which can lead to forms of matter with novel properties.

Nowadays, 5D relativity has a large number of adherents from both the physical and mathematical sides, those who are attracted by the elegance of matter from geometry. A significant group of these espouse Membrane theory as opposed to Space-Time-Matter theory. This was introduced in 1998 / 1999 by Arkani-Hamed / Dimopoulos / Dvali and Randall / Sundrum. The approach is typified by a singular surface or membrane which is identified as spacetime, whereas STM treats all of the five dimensions on an equal footing (though a divergence can appear under certain circumstances in the properties of the vacuum which may be related to the big bang). However, it has been shown that mathematically STM-theory and M-theory are equivalent, so to this extent they are complementary.

The present book aims to give a comprehensive account of 5D relativity, concentrating on how matter is induced in 4D spacetime by the fifth dimension. It effectively supersedes my two previous volumes on the subject, namely *Space-Time-Matter* of 1999 and *Five-Dimensional Physics* of 2006. Some of the material in these earlier volumes is also present in the current one, but there is also much that is new. Most of the new results are due to my colleagues, whose names are found within. I am grateful for their assistance, but of course the responsibility for any errors or omissions rests with me. The coverage is by chapters which I have attempted to make reasonably self-contained (each has its own bibliography for ease of reference). It starts with an overview of standard general relativity, proceeds through established results on astrophysics and cosmology, and ends with recent material on the nature of the vacuum and particle physics. The subject matter is perforce mathematical, each chapter has an introduction and conclusion of a non-technical nature aimed at maximizing understanding.

While the idea of matter from geometry goes back to Einstein, it falls into the kind of philosophy of science promoted by his colleague, Sir Arthur Eddington. He was Plumian Professor of Astronomy and Natural Philosophy at Cambridge, and I was influenced by reading his books while a graduate student at the same place, and acquired a rather liberal view of what should be considered "real". For example, one could argue that the planet Neptune should have been considered "real" before Hershel's eye saw it through a telescope, because the perturbations of the other planets considered within the framework of Newton's laws of mechanics *required* its existence. And in a more modern vein: gravitational radiation should be considered real even though no human eye has seen it, because the evolution of the binary pulsar considered within the framework of Einstein's general relativity requires its existence. I mention these things because they are relevant to the question I am frequently asked: "If the fifth dimension is real, why do we not see it?" The reader is naturally free to regard the fifth dimension in the Space-Time-Matter theory as he or she pleases, and at one end of the philosophical spectrum it can be viewed merely as a mathematical abstraction that gives convenient physical results. However, I would venture to answer the question posed above in a bolder way: in a manner, we see the fifth dimension all around us, in the form of mass and energy.

Paul S. Wesson
Gabriola Island, September 2015

Preface
by Second Author

Notes for this book were found among Paul Wesson's papers after his death on September 16, 2015. With the kind encouragement of Paul's partner, Patricia Lapcevic, I undertook to see that they were published. Chapters 1 through 8 are largely as Paul wrote them, with only minor editorial changes and typographical corrections. (Many thanks to Jonathan Perry for help in reformatting the equations.) I have added a concluding Chapter 9 and an Appendix describing Paul Wesson's life and work as I knew it. Any errors and omissions in this additional material are entirely my own. I was fortunate to know Paul as a teacher, colleague and friend, and I hope that this book, his last work, will be a fitting tribute to a life dedicated to the search for truth and beauty in physics.

James M. Overduin
Baltimore, September 2018

Contents

Chapter 1

Standard General Relativity

1.1. Introduction

The fifth dimension in modern physics is used to explain matter, be it the intensity of diffuse starlight in the universe or the mass of a particle in the laboratory. The majority of this book is concerned with recent developments in this subject, though it actually goes back to Einstein. As noted elsewhere, he dreamed of transmuting the "base wood" of matter into the "fine marble" of geometry. He did not succeed in doing this during his lifetime, but we know now that it is indeed possible to obtain a unified theory of the gravitational field and its source matter, provided we augment spacetime with an extra dimension.

Space-Time-Matter theory with five dimensions is, however, based on Einstein's theory of general relativity with four dimensions; and to properly understand the former it is necessary to have a working knowledge of the latter. To this end, the present chapter gives a compact, pedagogical account of gravity and matter in four-dimensional (4D) spacetime, while the rest of this book deals with the nature of things in 5D and higher. Those readers conversant with Einstein's theory may safely levitate from here to the material on the fifth dimension which succeeds it. However, the present chapter contains material which may be of interest even to the expert, including comments on modern observations and how well (or not) they agree with theory.

Einstein's theory has a vast literature, but those who do work with it tend to gravitate to a few books (some of which are massive enough to justify the metaphor). The ones in the bibliography have different strengths, and together cover everything that is necessary for an understanding of the basics of the theory [1]. There are also certain subjects which occur in the following sections that are discussed in books and papers of a more technical sort [2-8]. Ideas at the forefront of research are of a diverse nature, and perhaps best approached through introductory accounts [9-11]. Here, we review the essentials of Einstein's equations, cosmology and astrophysics in Sections 1.2, 1.3 and 1.4 respectively.

1.2. Einstein's Equations

This section is devoted to the genesis and properties of the field equations. The notation is standard, so $x^{0,123}$ are the coordinates of time and ordinary space. To avoid symbolic clutter, we will usually adopt the ploy of imagining that we measure time, distance and mass in units which make the speed of light c, Newton's constant of gravity G and Planck's constant of action h all equal to unity.

The so-called fundamental constants are, in fact, not very significant in their scientific content and are only constants in the sense of being useful conventions. They arise because the history of physics saw it useful to separate the things it deals with into categories, which in mechanics we label mass, length and time [6, 9, 10]. We ascribe basic units to these things, denoted in the abstract by M, L, T and in practice by convenient measures like the gram, centimetre and second. The latter are obviously man-made, but so are the former. The concepts of mass, length and time are instructive, and arise because of the ways in which humans perceive the world and comprehend it by the five senses. Over centuries of research, this approach has been honed, and nowadays we take it for granted that the equations of physics should be homogeneous in their physical dimensions.

Dimensional analysis – the traditional shortcut of the physicist – is really the application of an elementary form of group theory related to the Pi Theorem. It provides a way of checking the dimensional consistency of the equations of physics under the permutations of three base quantities

M, L, T. Dimensional analysis does not, of course, determine the dimensionless factors which may enter a problem, such as π or e. In this regard, it should be noted that the constants of physics *do* serve the useful purpose of converting a physical proportionality to an equation in numbers. To illustrate, let us consider the classical Kepler problem. In it, the Earth (mass m) orbits the Sun (mass M) with an azimuthal velocity (v) at a certain radial distance (r). The relative motion of the frames of reference of the two objects results in what historically has come to be called the centrifugal force mv^2/r. This is counterbalanced by the gravitational force of attraction between the objects, which following Newton we know to be proportional to the product of the masses and the inverse square of their separation. The essential physics of the Kepler orbit is described by the proportionality $mv^2/r \propto Mm/r^2$. However, to convert this to an *equation* we have to insert on the right-hand side an appropriate constant G. Its purpose is to transpose the physical characteristics of the quantities on the one side of the law to those on the other side, so it perforce has the physical dimensions of $M^{-1}L^3T^{-2}$. It is the somewhat arbitrary manner in which constants like G are introduced that has led several well-known workers to regard their presence in physics as accidental. By contrast, the cancelling of the m on the left-hand side of the previous relation with the m on the right-hand side is *not* trivial. It is a consequence of Einstein's Equivalence Principle, to which we will return below. The simplicity of the Kepler problem, and particularly of the answer $v = \sqrt{GM/r}$, is due to this Principle. Indeed, it is the fact that its laws are independent of the mass of a test object which makes gravitation a relatively simple science.

Quantum mechanics, in distinction to gravitation, is characterized by the unit of action h introduced by Planck and named after him. Both branches of science make use of c, the speed of light in vacuum. The complete suite of constants with their physical dimensions is thus $G = M^{-1}L^3T^{-2}$, $h = ML^2T^{-1}$, $c = LT^{-1}$. While these constants are commonplace, it is important to realize that their dimensional contents do not 'overlap': each may be set to unity by an appropriate choice of units independent of the others.

A corollary of this is that the mass of an object m can be geometrized, if so desired, in both subjects. The appropriate lengths are

$$Gm/c^2 \quad \text{and} \quad h/mc \ ,$$

the Schwarzschild radius and the Compton wavelength respectively. The existence of these implies that it is possible, at least in principle, to construct a unified theory of gravitation and the interactions of particle physics which is based on geometry.

It is also possible, as realized long ago by Planck and others, to use G, h and c to define a 'natural' set of units. (It is currently more common to use angular frequency than straight frequency in atomic problems, so $\hbar \equiv h/2\pi$ is the preferred unit.) The correspondence between natural or Planck units and conventional SI units can be summarized as follows:

$$1 \, m_p \equiv \left(\frac{\hbar c}{G}\right)^{\frac{1}{2}} = 2.2 \times 10^{-8} \text{ kg} \quad , \quad 1 \text{ kg} = 4.6 \times 10^7 m_p \ ,$$

$$1 \, \ell_p \equiv \left(\frac{G\hbar}{c^3}\right)^{\frac{1}{2}} = 1.6 \times 10^{-35} \text{ m} \quad , \quad 1 \text{ m} = 6.3 \times 10^{34} \ell_p \ ,$$

$$1 \, t_p \equiv \left(\frac{G\hbar}{c^5}\right)^{\frac{1}{2}} = 5.4 \times 10^{-44} \text{ s} \quad , \quad 1 \text{ s} = 1.9 \times 10^{43} t_p \ .$$

In Planck units, all the constants $G, \hbar$ and c become unity and they consequently disappear from the equations of physics.

In general relativity, the masses of objects are nearly always taken to be constants. It is therefore a theory of accelerations rather than forces. The Equivalence Principle, noted above, thus states that test masses accelerate in a gravitational field at the same rate, irrespective of their composition. This refers not only to chemical composition, but also to contributions to effective mass from binding energy and electromagnetic and other types of energy. For a particle, the Equivalence Principle removes the distinction which might be made between the gravitational mass (the quantity concerned in the object's gravitational field) and the inertial mass (the quantity which measures the object's energy content). For a fluid, however, it will be seen below that this distinction still exists and indeed follows from the field equations. The latter should not, of course, lead to consequences which depend on our choice of coordinates.

The Principle of Covariance makes formal this arbitrariness of coordinate, and by use of tensors ensures that the theory leads to results whose context is independent of how we describe things. As in other theories, in general relativity the prime objective is often the calculation of the path of a test particle. The Geodesic Principle provides a formal scheme for doing this. The analog of the distance between two nearby points in the four dimensions of spacetime is the elemental interval ds, which also defines proper time. The interval can be extremized by varying it to isolate the shortest route, as in the symbolic relation $\delta[\int ds] = 0$. The result is the geodesic equation whose four components give the equations of motion along the time and spatial axes (the time component involves the energy while the components in ordinary 3D space involve the momenta of the test particle). The three principles outlined in this paragraph, to do with Equivalence, Covariance and the Geodesic, form the basis of a theory which is both monolithic and intricate.

Einstein's field equations are usually presented as a match between the gravitational field and its source in matter. Some of the philosophical implications of this are still under discussion (see below), but the mathematical structure of the theory is straightforward. The interval between two nearby points in spacetime is defined via an extension of Pythagoras' theorem by $ds^2 = g_{\alpha\beta}dx^\alpha dx^\beta$, where a repeated index upstairs and downstairs is shorthand for summation over time (x^0) and space (x^{123}). The metric tensor $g_{\alpha\beta}$ is a 4×4 array of potentials, which is taken to be symmetric and so has ten independent elements. Generally the potentials depend on space and time, $g_{\alpha\beta} = g_{\alpha\beta}(x^\gamma)$, but locally they are constants whose magnitudes may be set to unity, defining flat Minkowski spacetime where the diagonal components are $\eta_{\alpha\beta} = (+1, -1, -1, -1)$. Derivatives of $g_{\alpha\beta}$ with respect to the coordinates define the useful objects named after Christoffel, $\Gamma^\alpha_{\beta\gamma} = \frac{1}{2}g^{\alpha\delta}(\partial_\gamma g_{\beta\delta} + \partial_\beta g_{\gamma\delta} - \partial_\delta g_{\beta\gamma})$. Here the symbol '$\partial_\alpha$' denotes the partial derivative with respect to x^α. This should not be confused with the corresponding covariant derivative, denoted by '∇_α' which takes into account the curvature of spacetime. (The covariant derivative of a vector V_α, for example, is given by $\nabla_\beta V_\alpha = \partial_\beta V_\alpha - \Gamma^\gamma_{\alpha\beta}V_\gamma$.) The Christoffel symbols figure in the geodesic equation mentioned above, which specifies the acceleration of a test particle in terms of its 4-velocity $u^\alpha \equiv dx^\alpha/ds$, via $du^\gamma/ds + \Gamma^\gamma_{\alpha\beta}u^\alpha u^\beta = 0$. They are also used to define the Riemann tensor $R^\alpha_{\beta\gamma\delta}$, which encodes all of the relevant information about the gravitational field. However, the Riemann tensor has 20 independent components, whereas to obtain field equations to solve for the 10 elements of the metric tensor $g_{\alpha\beta}$ requires an object with the same number of components. This is provided by setting the upper index in $R^\alpha_{\beta\gamma\delta}$ equal to one of the lower indices, and summing, a process which produces the contracted tensor $R_{\mu\nu}$ named after Ricci. When this is again contracted by taking its product with the metric tensor in its 'upstairs' or contravariant form, the result is $R = g^{\mu\nu}R_{\mu\nu} = R^0_0 + R^1_1 + R^2_2 + R^3_3$, the Ricci or curvature scalar. It can be thought of as a kind of measure of the average intensity of the gravitational field at a point in spacetime. Lastly, the combination $G_{\mu\nu} \equiv R_{\mu\nu} - (R/2)g_{\mu\nu}$ is of special interest because its 4D covariant divergence is zero by construction: $\nabla_\mu G^\mu_\nu = 0$. The geometrical object $G_{\mu\nu}$ is known as the Einstein tensor, and comprises the left-hand side of the field equations.

The preceding paragraph is standard material and familiar to many workers. However, it is not so widely known that Einstein wished to follow the same procedure for the *other* side of his field equations. That is, he wished to replace the common properties of matter, such as the density ρ and pressure p, by geometrical expressions. He termed the former "base wood" and the latter "fine marble". In the 1930s, Einstein attempted to find a way to effect this transmutation by using an extra dimension. This had already been shown by Kaluza to unify the gravitational and electromagnetic equations of classical theory, and Kaluza suggested an extension to quantum theory that was designed to explain the magnitude of the electron charge in terms of the momentum in the fifth dimension. Unfortunately, to make algebraic progress, Kaluza was obliged to assume that the 5D theory had functions independent of the fifth coordinate (the 'cylinder' condition), and Klein took the extra dimension to be rolled up to an unobservably small size ('compact-ification'). These two conditions proved to be a mathematical straitjacket for the theory, which robbed it of much of its physical power and doomed Einstein's dream of a unified account of gravity and matter. It was not until 1992 that a fully general 5D theory was formulated, which explained matter as being induced in 4D by the fifth dimension. As noted in Chapter

2, this theory was devised by workers trying to find a geometric rationale for rest mass, who were originally ignorant of Einstein's forgotten 'dream' (see also Chapter 9 and the Appendix for a historical account). Later, however, the rediscovery of an old embedding theorem of differential geometry due to Campbell showed that the 5D theory (based on the 5D Ricci tensor R_{AB}) contained the 4D one (based on the 4D Einstein tensor G_{AB}). This approach, known as Space-Time-Matter theory, was joined in 1998 by the similar Membrane theory; and it is now acknowledged that matter can be explained in geometric terms if so desired.

General relativity, in its regular 4D form, matches the Einstein tensor $G_{\alpha\beta}$ to an object which contains the phenomenological properties of matter, the energy-momentum tensor $T_{\alpha\beta}$. The form of this depends somewhat on the type of matter involved, but the latter is commonly assumed to be a perfect fluid (with an isotropic pressure and a unique density and no viscosity). Then the appropriate matter tensor may be written $T_{\mu\nu} = (\rho + p)u_\mu u_\nu - pg_{\mu\nu}$, where u_μ are the 4-velocities defined before. This form may look contrived, but it can be shown that the divergence $\nabla_\mu T_\nu^\mu = 0$ gives back the standard equations of motion in ordinary 3D space plus the equation of continuity (conservation of mass) for the fluid.

Before joining the parts of Einstein's equations which describe the gravitational field $(G_{\alpha\beta})$ and matter $(T_{\alpha\beta})$, it is necessary to tackle the notorious problem posed by the cosmological constant Λ [11]. The mathematical possibility of adding a term $\Lambda g_{\mu\nu}$ to the field equations arises because the metric tensor acts like a constant under covariant differentiation $(\nabla_\gamma g_{\alpha\beta} = 0)$. The presence of such a term does not therefore upset the physical considerations used to identify the left-hand side $(G_{\mu\nu})$ and the right-hand side $(T_{\mu\nu})$ of the proposed field equations. Notwithstanding this, it *does* have physical consequences. Notably, in a 3D spherically-symmetric distribution of matter, an acceleration appears which at radius r is $\Lambda r/3$. This is a repulsion for $\Lambda > 0$, but an attraction that augments gravity if $\Lambda < 0$. Einstein strongly disliked such a Λ-term, because it acts on matter without being itself connected with matter. But Eddington, his contemporary, regarded the Λ-term as an essential foundation of cosmology, and present observations do indeed indicate its importance (see below).

There has been much hand-wringing about the cosmological constant, both in physics and philosophy [11]. It continues to be a subject of controversy, because certain models of elementary particles imply intense vacuum fields which correspond to a large magnitude for Λ, in apparent contradiction with astrophysical observations which imply a small, positive value for Λ of order 10^{-56} cm^{-2}. The apparent discrepancy between theory and observation amounts to some 120 orders of magnitude. One reasonable way of explaining this is in terms of a 5D theory where Λ varies with scale depending on the size of the extra dimension (which, though, is controversial).

A new way to approach the cosmological constant may be to take from particle physics the idea that the vacuum is not merely emptiness but the seat of significant physics, and join this to the structure necessary for a tensor-based description of gravity like general relativity. The result is that Λ may be regarded as measuring the density and pressure of the vacuum, its equation of state being $\rho_v = -p_v = +\Lambda/8\pi$. This is neat, but not without its pitfalls. For example, it is common to take the physical dimension of Λ as L^{-2}, so with conventional units restored the dimensionally-correct form of the density is $\rho_v = \Lambda c^2/8\pi G$. This gives the impression that the vacuum is ultimately related to the strength of gravity, as measured by G. However, this is mistaken. Firstly, because there is a coupling constant $8\pi G/c^2$ in front of the energy-momentum tensor if the field equations are set up using conventional units, and this exactly cancels the similar factor in ρ_v as written above. Secondly, the so-called fundamental constants are in fact disposable, as we saw before; and while it may be convenient to put them back at the end of a complicated calculation, the numerical size of a given constant depends on an arbitrary choice of units and has no real significance. By contrast, the geometrical factor in $\rho_v = \Lambda c^2/8\pi G$ does have significance. It is composed of a conventional factor 2 connected with the standard way of expressing potentials, and a factor 4π. This is connected with the fact that the surface area of a sphere of radius r around a given centre in flat space is $4\pi r^2$, so the intensity of a conserved field necessarily falls off as $1/4\pi r^2$, and it is necessary to integrate over the same surface area in order to evaluate the strength of a source. This situation is identical to the one in classical electromagnetism as described by Maxwell's equations. Those equations are vectorial in nature, and admit of a gauge term which is the gradient of a scalar function. Similarly, while Einstein's equations are tensorial in nature, they too admit of a kind of gauge term. This is just the $\Lambda g_{\mu\nu}$

discussed above. In other words, the most satisfactory way to regard the cosmological constant is as a kind of gauge term for the equations of general relativity.

Putting Einstein's field equations together is now – in light of the foregoing discussion – a straightforward business. We choose to keep the Λ term explicit, both for mathematical generality and because it is indicated by modern observations. The equations in standard form then read

$$G_{\mu\nu} + \Lambda g_{\mu\nu} = 8\pi T_{\mu\nu} \ . \tag{1.1}$$

These equations, despite occupying only one line, entail a vast amount of physics. They are also remarkable in that they attempt to explain reality (as expressed by $T_{\mu\nu}$) in terms of a purely abstract quantity based on geometry (namely $G_{\mu\nu}$). While a precursor may be found in Maxwell's theory, Einstein's theory represents a fundamental break with older, mechanical ways of viewing the world. It is not the purpose of the present account to go into the many observations and tests which support the validity of Eqs. (1.1). But given the abstract mode of their genesis, it is truly remarkable that they work.

1.3. 4D Cosmology

In this section, the usual viewpoint is adopted that the universe started in some event like a big bang. This is indicated by the traditional evidence to do with cosmic redshift (expansion), the cosmic microwave background (CMB), light element synthesis, and the origin and growth of large-scale structure.

To these should be added the more recent evidence of the extragalactic background light (EBL) produced by sources like stars in galaxies [3]. This light is dependent on astrophysical processes, and should not be confused with the cooled-down fireball radiation now making up the CMB. The EBL has a very low intensity, which in the optical band has only recently been constrained in a meaningful way, and is controlled by the intensity of the sources, the redshift effect of the Hubble expansion, and the age of the universe. The last factor is important, and models of the integrated radiation from galaxies confirm that the present age is $t_0 \approx 14 \times 10^9$ yr approximately. The night sky is dark mostly because the universe is young, implying that something like a big bang occurred not too much longer ago than the lifetimes of 'typical' stars like the Sun.

The present universe, on the basis of observations of supernovae magnitudes and redshifts, appears to be accelerating under the influence of the cosmological constant or some similar scalar field, known generically as 'dark energy'. There is also ample evidence from the structure of spiral galaxies, the morphology of clusters of galaxies, and the gravitational lensing of distant sources like quasars, that there is a significant density of dark matter in the universe. The nature of this is controversial, but it could be elementary particles of some kind with a low effective temperature. Ordinary matter, of the kind seen in stars and the optical parts of galaxies, makes up a relatively small fraction of the whole, especially in comparison to the effect of the cosmological constant regarded as vacuum energy (Section 1.2 above). The relative densities of the vacuum, dark matter and ordinary matter are 73% : 23% : 4% approximately. And even the 'ordinary' or baryonic matter referred to here is mostly dark, in that it neither gives off nor absorbs significant amounts of electromagnetic radiation. The stuff of traditional astronomy is a mere smattering.

It is difficult to match the aforementioned data to any simple model of cosmology. It is particularly challenging to find a single set of parameters which describes the evolution of the scale factor $R(t)$ as a function of cosmic time over its entire history. For this reason, the current picture is largely qualitative. Following the big bang, there appears to have been a brief phase of extremely rapid expansion or 'inflation,' with the equation of state of the vacuum ($p = -\rho$), when the universe became relatively smooth. Then there was a hot period as the primeval fireball expanded more slowly and the equation of state of the matter was close to that of radiation ($p = \rho/3$). This evolved with cooling into the phase dominated by cold, dust-like matter ($p \approx 0$) and characterized by the appearance and growth of large-scale structure. Finally, as indicated by supernovae observations as well as other data, a transition occurred into a late-time accelerating regime that is again dominated by Λ-like vacuum energy. To model these different phases, we need to find relevant solutions to the field equations (1.1).

The required solutions are named after Friedmann, Lemaître, Robertson and Walker (FLRW, often shortened unjustly to FRW). The first two reduced the field equations to a pair of convenient differential equations which will be examined below. The latter two workers isolated the corresponding form of the interval, which is useful for calculating distances and related quantities. The 4D interval consists of two parts: a simple time, and a measure for the 3D distance whose form ensures that all pieces are equivalent. The Robertson-Walker interval is given by

$$ds^2 = dt^2 - \frac{R^2(t)}{(1+kr^2/4)^2}(dr^2 + r^2 d\Omega^2) \quad . \tag{1.2}$$

Here $d\Omega^2 \equiv (d\theta^2 + \sin^2\theta \, d\varphi^2)$ defines the angular part of the metric in spherical polar coordinates. The radial part is expressed for ease in terms of a measure that is chosen to be comoving with the matter, which means that r in Eq. (1.2) is merely a distance *label*, the same at all time for a given galaxy. The 'actual' (changing) distance involves the scale factor $R(t)$, which measures the separation between two typical galaxies at time t. The rate of expansion is given by Hubble's parameter $H \equiv \dot{R}/R$, where an overdot denotes the total derivative with respect to time. The second derivative of $R(t)$ is sometimes quantified in terms of a deceleration parameter, $q \equiv -\ddot{R}R/\dot{R}^2$. This is dimensionless, while H has units of inverse time. (The present value of H is about 70 km/s/Mpc in terms of its traditional but rather perverse unit, and galaxies that are not too distant recede at velocities proportional to this and the distance.) The constant k in Eq. (1.2) is a normalized measure of the curvature of ordinary 3D space, and can be positive, negative or zero. It should be noted that an alternative form of Eq. (1.2) appears in some texts, obtained from it by a change in the radial coordinate, thus:

$$ds^2 = dt^2 - R^2(t)\left(\frac{dr^2}{1-kr^2} + r^2 d\Omega^2\right) \quad . \tag{1.3}$$

This is useful if we wish to measure r relative to ourselves considered as 'centre', whereas Eq. (1.2) is spatially isotropic and provides a more 'global' measure. Of course, for both forms, there is no real centre and no boundary.

When the Robertson-Walker interval is used in conjunction with the Einstein field equations (1.1), the latter take the form of two relations which were studied by Friedmann and Lemaître. The assumption that the density ρ and pressure p of the cosmological fluid are isotropic and homogeneous (= uniform) reduces the partial differential equations (1.1) to ordinary differential equations for the scale factor $R(t)$ which measures the expansion. The Friedmann-Lemaître equations are

$$8\pi\rho = \frac{3k}{R^2} + \frac{3\dot{R}^2}{R^2} - \Lambda \quad , \tag{1.4a}$$

$$8\pi p = \frac{-k}{R^2} - \frac{\dot{R}^2}{R^2} - \frac{2\ddot{R}}{R} + \Lambda \quad . \tag{1.4b}$$

The curvature constant k here is usually normalized by choice of units for R to have the values ± 1, 0. It can be thought of as indicating the relative contributions of the kinetic energy and gravitational binding energy for a unit volume of the fluid. In the absence of Λ, $k = -1$ means that the balance of energies is in the direction of continued expansion, $k = +1$ means that the fluid eventually stops expanding and collapses under its own gravity, while $k = 0$ means an exact balance with a continuing but slowing expansion. However, Λ is not absent in the real universe, which considerably complicates the dynamical solutions of Eqs. (1.4), most of which can only be isolated by numerical means.

Some instructive things emerge from the two Friedmann-Lemaître equations (1.4) when they are combined in appropriate ways. For this, it is useful to replace Λ by its equivalent vacuum properties (see above), and write the total density and pressure as $\rho = \rho_m + \rho_v$, $p = p_m + p_v$ with matter and vacuum parts. Combining Eq. (1.4a) with three times Eq. (1.4b) to eliminate k then gives

$$\ddot{R} = \frac{-(4/3)\pi R^3(\rho+3p)}{R^2} \quad . \tag{1.5}$$

This is seen to be a quasi-Newtonian law of inverse-square attraction, when we recall that the physical distance in 3D is proportional at any time to the scale factor $R(t)$, though this symbol does not imply a physical boundary since the cosmological fluid is continuous. It is noteworthy that the effective gravitational mass of a portion of the fluid is proportional to the combination $(\rho + 3p)$, not the Newtonian ρ (which is only recovered for $p \ll \rho$). Accordingly, the combination $(\rho + 3p)$ is called the gravitational density. For pure vacuum, this combination is negative for $\Lambda > 0$ since $p = -\rho = -\Lambda/8\pi$, and this is why a universe dominated by a positive cosmological constant experiences a cosmic repulsion. Another instructive thing emerges when the first derivative of Eq. (1.4a) is combined with Eq. (1.4b) to eliminate $\ddot{R}$, giving:

$$\dot{\rho} = -3(\rho + p)\frac{\dot{R}}{R} \ . \tag{1.6}$$

This is seen to be a kind of stability relation for the universe, in the sense that the density adjusts in proportion to the expansion rate and the combination $(\rho + p)$. This is not a gravitational effect, and accordingly the noted combination is called the inertial density. For pure vacuum, the combination $(\rho + p)$ is zero since the equation of state is $p = -\rho$. So the vacuum has constant density (and pressure) even though the matter in the universe is expanding.

It is apparent from the above that the universe according to Einstein can have properties quite different from those imagined by Newton. The reasons for this have primarily to do with the cosmological 'constant', the possibility that the pressure of matter may be a significant fraction of the energy density, and the fact that the speed of light is large but finite. The last of these has consequences which are subtle but ubiquitous. To briefly review these, let us temporarily reinstate conventional (non-geometrical) units for c. Then it is obvious that as we look to greater distances we also look back in time. Advances in observational techniques are such that we can soon expect to be able to study in detail the first generation of galaxies. At greater distances we would 'see' the primordial plasma from which the galaxies formed. At still greater distances, we might even imagine witnessing the big-bang fireball itself. Since the universe is isotropic about every point, including us, this would resemble a glowing shell in every direction. We do in fact see such a shell, but it consists of photons that were emitted several hundred thousand years after the big bang. We cannot see farther, even in principle, because the universe at earlier times was so hot and dense that matter and light were in thermal equilibrium, making the universe opaque. By the time they reach us in the form of the cosmic microwave background or CMB, photons from this 'last scattering surface' have had their wavelengths stretched and energies diluted by cosmic expansion. Their spectrum, however, is still that of a perfect blackbody, reflecting their origin in the furnace of the big bang.

　　Horizons of a more fundamental kind also block our view of the remote cosmos, as they do our view of the distant parts of the Earth [1]. In the cosmological context, there are actually two kinds of horizon: an 'event horizon' separates those galaxies we can see from those we cannot ever see even as $t \to \infty$; while a 'particle horizon' separates those galaxies we can see from those we cannot see now at $t = t_0 \approx 14 \times 10^9$ yr. Cosmological models exist which have both kinds of horizon, one but not the other, or neither. To investigate these, consider the path of a photon which moves radially through a universe whose distances are defined by the Robertson-Walker metric. We put $ds = 0, d\theta = d\varphi = 0$ in Eq. (1.3) and obtain the (coordinate-based) velocity as $dr/dt = \pm c\sqrt{1 - kr^2}/R(t)$. The sign choice here corresponds to whether the photon is moving towards or away from us. More importantly, we see that the 'speed' of the photon is *not* just c. It depends on $R(t)$, which is given by the Friedmann-Lemaître equations (1.4). This means that the distance to the particle horizon, which defines that part of the universe in causal communication with us, can be complicated to work out. However, algebraic expressions can be written down for the simple case where $\Lambda = 0$ and $p = 0$. Then the distances are given by:

$$d = \begin{cases} \dfrac{c}{H_0\sqrt{2q_0-1}}\cos^{-1}\left(\dfrac{1}{q_0} - 1\right), & k = +1, q_0 > \dfrac{1}{2} \\[2mm] \dfrac{2c}{H_0} = 3ct_0, & k = 0, q_0 = \dfrac{1}{2} \\[2mm] \dfrac{c}{H_0\sqrt{1-2q_0}}\cosh^{-1}\left(\dfrac{1}{q_0} - 1\right), & k = -1, q_0 < \dfrac{1}{2} \end{cases} \qquad (1.7)$$

The Hubble parameter and deceleration parameter used here were defined above and the subscript '0' indicates that they are to be evaluated at the present epoch. It is apparent from these relations that the size of that part of the universe we can see is *not* just given by the product of the speed of light and the age.

The redshift z is in some ways a better parameter to use as a cosmological measure than either the distance or the time. It is a quantity which is directly observable; and it runs smoothly from us ($z = 0$), through the populations of galaxies and quasars ($z \approx 1 - 10$) and in principle all the 'way' to the big bang ($z \to \infty$). It is defined in terms of the scale factor of the Robertson-Walker metric at present (t_0) and at emission (t_e) by $1 + z \equiv R(t_0)/R(t_e)$. This neatly sidesteps long-running arguments about whether the redshift is 'caused' by the Doppler effect, gravity or some other agency, which are frame-dependent in general relativity and cannot be uniquely identified. The noted definition merely makes a statement about light waves and a ratio of scales. (It might even be imagined that the universe is momentarily static at the two instants which define the redshift, with no information available as to what happened in between.) Notwithstanding the utility of the redshift as a measure, it is still true that most workers have a mental picture of a universe that evolves through stages separated in time. This is actually acceptable, provided the epoch is used only in a relative sense, as an ordering device. Let us therefore return to this mode of organization, and list the solutions of the Friedmann-Lemaître equations (1.4) relevant to the successive phases of the universe.

Inflation is characterized by a rapid expansion under the influence of the cosmological constant or some similar measure of vacuum energy. The appropriate solution of Eq. (1.4) was found by de Sitter in the early days of general relativity and is given by

$$p = -\rho = -\Lambda/8\pi \ , \quad R(t) \propto e^{t/L} \ , \quad k = 0 \ . \tag{1.8a}$$

The length scale here is related to the cosmological constant by $\Lambda = 3/L^2$ (the proportionality sign indicates that the scale factor is arbitrary up to a constant). The present universe also appears to have a significant value of

Λ, which corresponds to a length L of order 10^{28} cm. The interval corresponding to Eq. (1.8a) is

$$ds^2 = dt^2 - e^{2t/L}(dr^2 + r^2 d\Omega^2) \quad . \tag{1.8b}$$

There is an alternative form of this cosmological metric, which is related by a coordinate transformation but is local in nature, thus:

$$ds^2 = \left(1 - \frac{\Lambda r^2}{3}\right) dt^2 - \frac{dr^2}{(1 - \Lambda r^2/3)} - r^2 d\Omega^2 \quad . \tag{1.8c}$$

This form of the de Sitter metric has been extensively used to model quantum-mechanical processes in the early universe, like tunnelling. Such processes could be of great importance if it should be shown that general relativity needs to be extended in some way. For example, it then becomes feasible to explain the big bang as a quantum event, perhaps in a higher-dimensional manifold. In this regard, it can be mentioned that Eq. (1.8c) for both signs of Λ can be embedded in a 5D manifold which is *flat*, in which case Eq. (1.8c) resembles a 4D pseudosphere with radius L [8]. Similarly, Eq. (1.8b) can be embedded in 5D Minkowski space.

Following inflation, the universe is believed to have passed through a hot period when the matter had an equation of state similar to that of radiation. A solution of the Friedmann-Lemaître equations (1.4) with these properties has long been known, though it was formulated before the importance of Λ was realized. The formal solution has

$$p = \rho/3 = 1/32\,\pi t^2 \,, \quad R(t) \propto t^{1/2} \,, \quad k = 0 \,, \quad \Lambda = 0 \quad . \tag{1.9}$$

This solution needs to be modified as regards its global properties for $\Lambda \neq 0$, but its local properties are still those necessary for nucleosynthesis of the kind needed to explain the observed abundances of light elements.

Later, when the matter had cooled, the universe is believed to have evolved into a cold phase which persists to the present, and which is characterized by a value for the matter pressure which is effectively zero. The formal solution of Eq. (1.4) has

$$p = 0 \,, \quad \rho = 1/6\pi t^2 \,, \quad R(t) \propto t^{2/3} \,, \quad k = 0 \,, \quad \Lambda = 0 \quad . \tag{1.10}$$

This solution, like the previous one, needs to be modified in regard to its global properties for $\Lambda \neq 0$. The solution (1.10) is named after Einstein-de Sitter, and should not be confused with the straight de Sitter solution (1.8a). For many years, Eq. (1.10) was considered to be a good approximation to the real universe. Ironically, observation now indicates that the old solution (1.8a), with its cosmological constant first postulated and then discarded by Einstein as his 'biggest blunder,' comes considerably closer to the truth.

1.4. Astrophysics

The application of general relativity to astrophysical systems is simpler than to cosmology for one main reason: the influence of the cosmological constant is negligible. For this reason, we largely ignore it in this section. Also, despite what was stated in Section 1.2 about the disposability of the so-called fundamental constants, G and c are now made explicit in order to bring out the comparison with Newtonian theory and special relativity.

Many astrophysical systems are spherically symmetric in ordinary 3D space. The solar system is like this, though as a solution of Einstein's equations (1.1) it is exceptionally simple because it is approximately empty of matter except for the Sun (mass M). The interval may be regarded as an extended version of the local de Sitter one (1.8c), and is given by

$$ds^2 = \left(1 - \frac{2GM}{c^2 r} - \frac{\Lambda r^2}{3}\right) dt^2 - \frac{dr^2}{(1 - 2GM/c^2 r - \Lambda r^2/3)} - r^2 d\Omega^2 \quad . \quad (1.11)$$

This is the standard form, but it should be noted that the potential can be written $2G(M + M_v)/c^2 r$ where $M_v = (4/3)\pi r^3 \rho_v$ is the effective mass of the vacuum due to its equivalent density $\rho_v = \Lambda c^2/8\pi G$ (see Section 1.2). It should also be noted that while the local de Sitter solution (1.8c) can be embedded in flat 5D spacetime, the Schwarzschild-de Sitter solution (1.11) *cannot* be embedded in a flat space of less than six dimensions. The fact that Eq. (1.11) successfully accounts for the dynamics of the solar system and binary pulsars, thereby establishing the validity of general relativity, means also that any extra dimensions must play a minor role in much of astrophysics.

To study other astrophysical systems where there is substantial matter, we assume the latter to be a spherically-symmetric perfect fluid described by the scalars ρ and p for the density and pressure. It is convenient to express the interval in the form

$$ds^2 = e^\sigma c^2 dt^2 - e^\omega dr^2 - R^2 d\Omega^2 \quad . \tag{1.12}$$

Here σ and ω are metric coefficients which in general depend on the time t and a radial measure r. The latter can be chosen to be comoving with the matter [1]. The latter may flow either inwards or outwards, but an element of it then maintains the same radial label r (as in the Robertson-Walker metric of Section 1.3). By contrast, $R = R(t, r)$ is really another metric coefficient, and measures the dynamics of the fluid, though in such a way that $2\pi R$ is the circumference of a great circle around the centre of the distribution. With this setup, it is the inequality of r and R in Eq. (1.12) that characterizes the departure of ordinary 3D space from flatness due to the gravitational field of the fluid.

Given the interval (1.12), the question arises of how to write Einstein's equations (1.1) in the most informative manner. In many texts, they are written as long strings of symbols relating the derivatives of the metric coefficients σ, ω, R to the properties of matter ρ, p. For problems of the type being considered here, there will in general be four equations for the five unknowns. Therefore, one relation may be specified in order to balance things and hopefully find a solution (ways to do this are examined below). However, in such problems it is often useful to define a function which is first order in the derivatives as a *new* unknown, and rewrite the four *second*-order partial differential equations as five *first*-order ones [4]. For the current problem, it was found some while ago by Podurets and Misner and Sharp that the appropriate new function to define is a measure of the mass of the fluid interior to radius r at time t, that is $m = m(r, t)$. The upshot is a set of five first-order differential equations in three metric coefficients (σ, ω, R) and three properties of the matter (ρ, p, m). Not only does this improve the tractability of the algebra, it also (after some manipulation) leads to a set of equations whose physical significance is more transparent.

Writing the definition of the mass function as a relation with other quantities, the full set of field equations is:

$$\frac{2Gm}{c^2 R} = 1 + \frac{e^{-\sigma}}{c^2}\left(\frac{\partial R}{\partial t}\right)^2 - e^{-\omega}\left(\frac{\partial R}{\partial r}\right)^2 \quad , \tag{1.13a}$$

$$\frac{\partial m}{\partial t} = \frac{-4\pi p R^2}{c^2}\frac{\partial R}{\partial t} \quad , \tag{1.13b}$$

$$\frac{\partial m}{\partial r} = 4\pi\rho R^2 \frac{\partial R}{\partial r} \quad , \tag{1.13c}$$

$$\frac{\partial \sigma}{\partial r} = \frac{-2}{p+\rho c^2}\frac{\partial p}{\partial r} \quad , \tag{1.13d}$$

$$\frac{\partial \omega}{\partial t} = \frac{-2c^2}{p+\rho c^2}\frac{\partial p}{\partial t} - \frac{4}{R}\frac{\partial R}{\partial t} \quad . \tag{1.13e}$$

The first of these equations, experience shows, is usually the hardest to solve. But it is helpful to note that it involves a balance between the Schwarzschild-like gravitational potential $Gm/c^2 R$, the kinetic energy per unit mass of the fluid $(\partial R/\partial t)^2$, and a measure of the departure of ordinary space from flatness, or equivalently the binding energy per unit mass of the fluid stored in the gravitational field $(\partial R/\partial r)^2$. The second equation above is best interpreted from right to left. It says, loosely speaking, that the force due to the pressure p acting over a shell of area $4\pi R^2$ that moves at a velocity $\partial R/\partial t$ forms a quantity which in mechanics would be termed a rate of work or power, and that the mass of the fluid responds by changing at a rate consistent with Einstein's formula for the equivalent energy mc^2. The third equation would on integration give the usual Newtonian expression for the mass of a portion of the fluid $(m = 4\pi R^3/3)$ if the space were flat $(R = r)$; but since it is not, Eq. (1.13c) gives the corresponding differential form for the mass of the fluid as affected by its own gravitational field. The last two equations (1.13d) and (1.13e), relate the metric coefficients to the properties of the matter responsible for curving spacetime.

Solving Eqs. (1.13) can be achieved once an extra relation is specified which balances the number of equations and the number of unknowns. There are also numerous solutions in the literature which were found by more tedious means, and whose physical meanings may be elucidated by employing Eqs. (1.13). It would be redundant to list those solutions here, especially since reviews are available [1, 2]. The relations (1.13) have been applied to a wide range of problems, since they cover everything from the global cosmological fluid (the Friedmann-Lemaître equations are included) to tiny perturbations of it [4, 5]. Thus they lead to a more objective form of the Cosmological Principle, in which all intelligent observers judge the universe to be the same everywhere, not merely in terms of the density and pressure but in terms of *dimensionless* combinations of these and other parameters. While at the other end of the spectrum, they can be used to study the growth of material around a quantum seed to form a proto-galaxy. The equations in the form (1.13) are especially useful in understanding the behaviour of matter under extreme circumstances, such as when the pressure approaches the energy density and the velocity of sound approaches the speed of light. New solutions like this certainly await discovery.

There are also several ways to specify a condition which makes the set of equations (1.13) determinate. Some examples:

(a) An equation of state, $p = p(\rho)$, is the traditional approach. This is particularly efficacious if additional information is available about the microscopic state of the matter, for example from spectral observations of a real system.

(b) Boundary conditions, in the broad sense, can help to restrict the form of a solution. These may include continuity conditions on the metric tensor if there is a join to another solution; or physical conditions, such as ones on the pressure at the centre and periphery of a system.

(c) Morphological constraints, such as self-similarity. The latter technique is especially relevant to astrophysical systems, which often lack sharp boundaries or other scales [5]. A distribution completely free of scales may be described by defining a dimensionless combined variable (say ct/r), so enabling the problem to be posed in ordinary rather than partial differential equations, which are easier to solve. A distribution with a single scale may be tackled using a refinement of this technique, so that problems like phase changes which involve a change in size of a physical parameter can be treated.

The preceding options are not exhaustive, and there is always the alternative of numerical integration. However, due to the non-linearity of Einstein's equations, exact algebraic solutions are especially valuable. Motivated readers are invited to join the search.

1.5. Conclusion

General relativity is widely accepted as the true theory of gravity, and is consistent with observational evidence stretching from GPS navigation systems on Earth to the most remote supernovae and quasars (subject to the assumption of vast amounts of as-yet undetected dark matter and dark energy). Much of cosmology can be treated using the Friedmann-Lemaître equations (1.4) for a uniform fluid, and much of astrophysics can be handled by the more complicated equations (1.13) for a spherically-symmetric fluid. In these two areas, it remains to find a single model that describes the whole history of the universe, and solutions that do justice to the diversity of its constituent systems. Notwithstanding these technical shortcomings, it is still true to say that Einstein's theory provides a satisfactory account of the real universe.

There is, however, room for improvement in certain non-technical areas, and especially in what might be termed the psychology of cosmology. Anyone who has taught cosmology knows that even bright students have difficulty with the concepts raised by Einstein's theory. And even some researchers have an inadequate idea of what the big bang must have been like. This is largely because human beings are imprinted from childhood with everyday constructs which leave them ill-equipped as adults to visualize a universe without a centre or a boundary. Yet if the density and pressure depend only on time then logic tells us that neither thing can exist. Confusion is engendered by calling the big bang an explosion, because this brings to mind a conventional bomb that sends shrapnel out from a point in 3D space until it hits some obstruction like a wall. Insofar as an analogy can be made, the big bang should be imagined as a kind of explosion that fills all of 3D space at the same moment, as if there is an indefinitely large number of bombs which are wired together so that they all detonate at the same instant. Even this description does not capture all the subtleties of the Einstein singularity, and it is hoped that progress will be made in this area in the future.

Also in the future, we can expect a significant shift in our understanding of the classical universe if a way is found to unify it with the quantum theory of particle interactions. That a connection exists is already hinted by the cosmological-'constant' problem, wherein the energy density of the vacuum is observed to be small on macroscopic scales but inferred to be large on microscopic scales. This problem might disappear if the properties of the vacuum prove to be variable. But such a compromise would still entail significant changes to our current accounts of both cosmology and particle physics. In fact, most workers believe that new physics will inevitably emerge from a unification of our present classical and quantum theories.

Currently, the most widely studied route to unification is via extra dimensions. The basic extension is to 5D, as in Space-Time-Matter theory and Membrane theory, whose main concerns are with classical matter and particles, respectively. For cosmology, perhaps the main consequence of the fifth dimension is the realization that the 4D big bang is a kind of artefact, produced by an unfortunate choice of coordinates in a flat 5D manifold. More generally, 5D relativity is a unified theory of gravitation, electromagnetism and a scalar field. It is the classical analog of the quantum interactions of the spin-2 graviton, the spin-1 photon and a spin-0 scalaron. The last may be related to the standard-model Higgs boson, whose existence generates the rest masses of elementary particles such as quarks and leptons (though in the classical theory particle masses also involve shifts in spacetime along the fifth dimension).

As to the old question: Why do we not 'see' the extra dimension? Well, in a way we do: it is the mass/energy all around us. This may sound strange; but adding one or more extra dimension(s) is actually the most effective way to extend general relativity so as to obtain new physics without upsetting established knowledge. With Einstein's theory as a foundation, we proceed in the following chapters to see exactly how the fifth dimension works.

References

[1] Carroll, S., *Spacetime and Geometry: An Introduction to General Relativity* (Addison-Wesley, San Francisco, 2004); Rindler, W., *Relativity: Special, General and Cosmological* (Oxford Un. Press, Oxford, 2001); Islam, J.N., *An Introduction to Mathematical Cosmology* (Cambridge Un. Press, Cambridge, 1992); Misner, C.W., Thorne, K.S. and Wheeler, J.A., *Gravitation* (Freeman, San Francisco, 1973); Robertson, H.P. and Noonan, T.W., *Relativity and Cosmology* (Saunders, Philadelphia, 1968).

[2] Kramer, D., Stephani, H., MacCallum, M., Herlt, E., *Exact Solutions of Einstein's Field Equations* (Cambridge Un. Press, Cambridge, 1980).

[3] Overduin, J.M., Wesson, P.S., *The Light/Dark Universe* (World Scientific, Singapore, 2008); Overduin, J.M., Wesson, P.S., *Phys. Rep.* **402**, 267 (2004).

[4] Wesson, P.S., *Astron. Astrophys.* **68**, 131 (1978); Wesson, P.S., *J. Math. Phys.* **19**, 2283 (1978); Wesson, P.S., *J. Math. Phys.* **25**, 3297 (1984); Wesson, P.S. and Ponce de Leon, J., *Phys. Rev.* **D39**, 420 (1989); Wesson, P.S., *Astrophys. J.* **336**, 58 (1989).

[5] Henriksen, R.N., Emslie, A.G. and Wesson, P.S., *Phys. Rev.* **D27**, 1219 (1983); Wesson, P.S., *Phys. Rev.* **D34**, 3925 (1986); Wesson, P.S., *Astrophys. J.* **228**, 647 (1979).

[6] Petkov, V. (ed.), *Relativity and the Dimensionality of the World* (Springer, Dordrecht, 2007); Petkov, V. (ed.), *Minkowski Spacetime: A Hundred Years Later* (Springer, Dordrecht, 2010).

[7] Carr, B.J., *Universe or Multiverse?* (Cambridge Un. Press, Cambridge, 2007).

[8] Wesson, P.S., *Five-Dimensional Physics* (World Scientific, Singapore, 2006). Wesson, P.S., *Gen. Rel. Grav.* **40**, 1353 (2008); Wesson, P.S. and Ponce de Leon, J., *J. Math. Phys.* **33**, 3883 (1992).

[9] Halpern, P., *The Great Beyond: Higher Dimensions, Parallel Universes, and the Extraordinary Search for a Theory of Everything* (Wiley, Hoboken, 2004); Halpern, P., *The Pursuit of Destiny: A History of Prediction* (Perseus, Cambridge, 2000).

[10] Halpern, P. and Wesson, P.S., *Brave New Universe: Illuminating the Darkest Secrets of the Cosmos* (J. Henry, Washington, 2006); Wesson, P.S., *Weaving the Universe: Is Modern Cosmology Discovered Or Invented?* (World Scientific, Singapore, 2011).

[11] Kragh, H. and Overduin, J.M., *The Weight of the Vacuum: A Scientific History of Dark Energy* (Springer, Berlin, 2014).

Chapter 2

Space-Time-Matter Theory

2.1. Introduction

General relativity is a monolithic theory, and can reasonably be called complete as a theory of gravity in the four dimensions of spacetime. However, Einstein and others have argued that the gravitational field and its source in matter should not be regarded as separate things, but unified in some way. Also, of course, there is the ultimate goal of unifying gravity with the other interactions, something which will presumably entail an amalgam of classical and quantum physics. Since Einstein's field equations can be cast in any number of dimensions, an appealing result to a better theory is via extending spacetime, notably to 5D.

There is in principle no restriction on the number N of dimensions. Indeed $N \rightarrow \infty$ has been suggested, and might make physical sense if the world were organized in such a fashion that the higher dimensions contributed progressively smaller amounts of energy to a given process. However, what we know about the interactions indicates the opposite, and general relativity is often regarded as the low-energy section of a high-N theory in which the other dimensions are related to the symmetry groups exhibited by elementary particles. This implies that 5D is most likely to be worthy of study; and in fact its Riemannian geometry gives a natural framework for gravitation and electromagnetism, plus a scalar field whose properties are still under investigation.

Space-Time-Matter (STM) theory is a formalism in which the matter and energy that we see in 4D is a result of the geometry in 5D. In a way, matter in spacetime is induced by the fifth dimension.

There are other ways to employ the algebra of 5D Riemannian geometry, and in Section 2.2 we will consider alternative approaches, notably the original Kaluza-Klein theory and the more modern Membrane theory. However, in Section 2.3 we will see that the 5D field equations lead almost automatically to a 4D energy-momentum tensor which provides a good description of matter. Then in Section 2.4 some simple solutions of the field equations will be presented, whose properties confirm the STM approach. In Section 2.5 a short account will be given of Campbell's theorem, as it relates to general properties of the embedding of 4D in 5D and the correspondence between STM theory and membrane theory. It should be noted that the form of the field equations in Section 2.3 can be employed to hand-check the numerous solutions found in the rest of this volume, though they can also be validated much more quickly by computer.

2.2. Higher-Dimensional Theories

The idea that the world may have more than four dimensions is due to Kaluza, who in 1921 realized that a 5D manifold could be used to unify Einstein's theory of general relativity with Maxwell's theory of electromagnetism [1]. After some delay, Einstein endorsed the idea, but a major impetus was provided in 1926 by Klein. He made the connection to quantum theory by assuming that the extra dimension was microscopically small, with a size in fact connected via Planck's constant to the magnitude of the electron charge [2]. Despite its elegance, though, this version of Kaluza-Klein theory was largely eclipsed by the explosive development first of wave mechanics and then of quantum field theory. However, the development of particle physics led eventually to a resurgence of interest in higher-dimensional field theories as a means of unifying the long-range and short-range interactions of physics. The breadth of the subject means that reviews of it tend to be lengthy. In this section we will therefore abstract the main ideas and equations from some recent accounts [3-5], with a view to setting the scene for the introduction of the STM field equations in the following section.

Kaluza-Klein theory is essentially general relativity in 5D, but constrained by two conditions. Physically, both have the motivation of explaining why we perceive the four dimensions of spacetime and (apparently) do not see the fifth dimension. Mathematically, they are somewhat different, however. (a) The so-called 'cylinder' condition was introduced by Kaluza, and consists in setting all partial derivatives with respect to the fifth coordinate to zero. It is an extremely strong constraint that has to be applied at the outset of calculation. Its main virtue is that it reduces the algebraic complexity of the theory to a manageable level. (b) The condition of compactification was introduced by Klein, and consists in the assumption that the fifth dimension is not only small in size but has a closed topology (i.e. a circle if we are only considering one extra dimension). It is a constraint that may be applied retroactively to a solution. Its main virtue is that it introduces periodicity and allows the use of Fourier and other decompositions of the theory.

There are now fifteen dimensionless potentials, which are the independent elements in a symmetric 5×5 metric tensor g_{AB} ($A, B = 0,123,4$). The first four coordinates are those of spacetime, while the extra one $x^4 = \ell$ (say) is sometimes referred to as the "internal" coordinate in applications to particle physics. In perfect analogy with general relativity, one can form a 5D Ricci tensor R_{AB}, a 5D Ricci scalar R and a 5D Einstein tensor $G_{AB} \equiv R_{AB} - Rg_{AB}/2$. The field equations would logically be expected to be $G_{AB} = kT_{AB}$ with some appropriate coupling constant k and a 5D energy-momentum tensor. But the latter is unknown, so from the time of Kaluza and Klein onward much work has been done with the 'vacuum' or 'empty' form of the field equations $G_{AB} = 0$. Equivalently, the defining equations are

$$R_{AB} = 0 \quad (A, B = 0 - 4) \ . \tag{2.1}$$

These fifteen relations serve to determine g_{AB}, at least in principle.

In practice, this is impossible without some starting assumption about g_{AB}. This is usually connected with the physical situation being investigated. In gravitational problems, an assumption about $g_{AB}(x^C)$ is commonly called a choice of coordinates, while in particle physics it is commonly called a choice of gauge. We will meet numerous concrete examples later, where given the functional form of $g_{AB}(x^C)$ we will calculate the 5D analogs of the Christoffel symbols Γ^C_{AB} which then give the components of R_{AB}. Kaluza was interested in electromagnetism, and realized that g_{AB} can be expressed in a form that involves the 4-potential A_α that figures in Maxwell's theory. He adopted the cylinder condition noted above, but also put g_{44} = constant. We will do a general analysis of the electromagnetic problem in Chapter 6, but here we look at an intermediate case where $g_{AB} = g_{AB}(x^\alpha)$, $g_{44} = -\Phi^2(x^\alpha)$. The metric tensor in matrix form is then

$$g_{AB} = \begin{bmatrix} \left(g_{\alpha\beta} - \kappa^2\Phi^2 A_\alpha A_\beta\right) & -\kappa\Phi^2 A_\alpha \\ -\kappa\Phi^2 A_\beta & -\Phi^2 \end{bmatrix} , \tag{2.2}$$

where κ is a coupling constant. The field equations (2.1) reduce to

$$G_{\alpha\beta} = \frac{\kappa^2\Phi^2}{2} T_{\alpha\beta} - \frac{1}{\Phi}\left(\nabla_\alpha\nabla_\beta\Phi - g_{\alpha\beta}\Box\Phi\right) ,$$

$$\nabla^\alpha F_{\alpha\beta} = -3\frac{\nabla^\alpha\Phi}{\Phi} F_{\alpha\beta} , \tag{2.3}$$

$$\Box\Phi = -\frac{\kappa^2\Phi^3}{4} F_{\alpha\beta}F^{\alpha\beta} .$$

Here $G_{\alpha\beta}$ and $F_{\alpha\beta}$ are the usual 4D Einstein and Faraday tensors, and $T_{\alpha\beta}$ is the energy-momentum tensor for an electromagnetic field given by $T_{\alpha\beta} = \left(g_{\alpha\beta}F_{\gamma\delta}F^{\gamma\delta}/4 - F^\gamma_\alpha F_{\beta\gamma}\right)/2$. Also ∇_α is the 4D curved-space or covariant derivative, $\Box \equiv g^{\alpha\beta}\nabla_\alpha\nabla_\beta$ is the wave operator (D'Alembertian) and the summation convention is in effect. Therefore we recognize the middle member of Eqs. (2.3) as the four equations of electromagnetism modified by a function, which by the last member of Eqs. (2.3) can be thought of as depending on a wave-like scalar field. The first member of Eqs. (2.3) gives back the ten Einstein equations of 4D general relativity. However, the energy and momentum of the right-hand side are effectively derived from the fifth dimension. In short, Kaluza-Klein theory is a unified account of gravity, electromagnetism and a scalar field.

Kaluza's original case $g_{44} = -\Phi^2 = -1$ together with the identification $\kappa = (16\pi G/c^4)^{1/2}$ makes Eqs. (2.3) read

$$G_{\alpha\beta} = \frac{8\pi G}{c^4} T_{\alpha\beta} \;\; ,$$

$$\nabla^\alpha F_{\alpha\beta} = 0 \;\; .$$

(2.4)

These are of course the straight Einstein and Maxwell equations in 4D, but derived from vacuum in 5D, a consequence which is sometimes referred to as the Kaluza-Klein "miracle". However, these relations have no contribution from the scalar field. The latter could well be important, particularly in application to particle physics. In the language of that subject, the field equations (2.3) of Kaluza-Klein theory describe a spin-2 graviton, a spin-1 photon, and a spin-0 boson which is thought to be connected with the way in which particles acquire mass.

It should be emphasized, however, that the definition of particle mass in 5D theory is subtle. We will discuss one possible approach to this problem in Chapter 9; but some workers feel that rest mass cannot be properly treated within the framework provided by 5D, and needs a higher-D approach. A popular one is 10D supersymmetry, whose rationale is to remove the large vacuum fields that otherwise attend models of elementary particles by partnering every boson (integral spin) with a fermion (half-integral spin). Thus the hypothetical carrier of gravity in general relativity, the graviton, would be partnered with a gravitino. The theoretical justification for supersymmetry is that any solution in 4D, no matter how energetic, can be embedded in a 10D manifold that is flat and empty. Actually, the formalism works somewhat better when extended to 11D, which case is known as supergravity. However, the practical support for supersymmetry in any of its forms is weak, in that none of the putative fermions (like the graviton) has been seen in particle accelerators. Also, if such weakly-interacting massive particles (WIMPs) are unstable to decay, their contribution to the density of the universe must be relatively minor, from data on the brightness of the night sky at different wavelengths [6]. Reviews of $N > 5D$ field theory, including strings, may be found elsewhere [e.g. Ref. 4]; so here we return to $N = 5$, and focus on modern versions of Kaluza and Klein's theory which make significant use of the scalar field and the extra coordinate.

The two main versions of 5D relativity are Space-Time-Matter theory and Membrane theory. Both avoid the cylinder and compactification conditions of Kaluza-Klein outlined above. They are nevertheless in agreement with extant observational data, including the classical tests [7]. However, both theories reduce to general relativity in the 4D limit, so neither has come up with a striking new test. Even so, they have both developed large literatures since they were proposed. STM theory was formulated in 1992, and was initially developed by Wesson, Ponce de Leon, Mashhoon, Liu and Overduin [8-11]. M theory dates from 1998-1999, and is based on ideas from Randall and Sundrum and Arkani-Hamed, Dimopoulos and Dvali [12-15]. Mathematically, the theories are similar, as shown by Ponce de Leon [16]. Conceptually, they differ significantly, as we will see.

Membrane theory is based on 5D manifold in which there is a singular hypersurface which we call 4D spacetime. It is motivated by the wish to explain the apparently weak strength of gravity as compared to the forces of particle physics. It does this by assuming the gravity propagates freely (into the 5D bulk), whereas particle interactions are constrained to the hypersurface (the 4D brane). The theory is also motivated by the wish to explain the relatively small masses of elementary particles, something which is known as the hierarchy problem. It also helps to explain the existence and size of the cosmological constant Λ (see Chapter 1), since this parameter sets the scale for distances measured away from the membrane. The 5D metric for M theory has the general form

$$dS^2 = e^{F(\ell)} g_{\alpha\beta} dx^\alpha dx^\beta \pm d\ell^2 \ . \tag{2.5}$$

This is the so-called warp metric, where the factor $F(\ell)$ is rendered dimensionless through its dependence on Λ. The membrane, in the original version of M theory at least, is located at $x^4 = \ell = 0$. (Most papers on M theory actually use $x^4 = y$, but this can be confused with the usual Euclidian coordinate, so we use instead the terminology of STM theory.) Physics is assumed to be symmetric above the membrane, so it has what is called Z_2 symmetry. This and other aspects of the theory have evolved somewhat, so there has been discussion of thick branes, the existence of singular or thin branes in $(4 + d)$ dimensions or d-branes, and the possible collisions of branes as a means of explaining the big bang of traditional 4D cosmology.

Space-Time-Matter theory is based on an unrestricted 5D manifold, in which the extra dimension serves to induce matter in 4D spacetime. It does this because the apparently empty 5D field equations $R_{AB} = 0$ contain the Einstein equations $G_{\alpha\beta} = 8\pi T_{\alpha\beta}$, with an effective energy-momentum tensor which depends on derivatives of the metric coefficients with respect to the extra coordinate $x^4 = \ell$. This is actually guaranteed by Campbell's embedding theorem of differential geometry, which will be summarized in Section 2.5. It is a theory where matter is derived from geometry. Or alternatively, a theory where the fields are united with their sources. This fulfills an old dream of Einstein, though the modern theory has extensive ramifications. These are the subjects of the other parts of this volume, so here we only mention the main aspects of the theory. One of these concerns the equations of motion, which by analogy with the empty nature of the 5D field equations are based on the concept of a 5D null path. It may in fact be shown that the 5D interval $dS^2 = 0$ contains the 4D ones with $ds^2 \geq 0$. So all particles are photon-like in 5D, though the massive ones travel on timelike paths in 4D. Another aspect of STM theory concerns the cosmological constant Λ. The 5D field equations are free of scales and do not contain such a parameter (unlike the equations of M theory, which sometimes include such a 5D term). However, the Λ of Einstein's theory emerges automatically from the equations when the metric takes on a particularly simple form:

$$dS^2 = \left(\frac{\ell}{L}\right)^2 g_{\alpha\beta}dx^\alpha dx^\beta \pm d\ell^2 \quad . \tag{2.6}$$

This is the so-called canonical metric, where the quadratic factor contains a constant scale which turns out to be fixed by $\Lambda = \mp 3/L^2$. This is so, provided the metric coefficients are independent of the extra coordinate, $g_{\alpha\beta} = g_{\alpha\beta}(x^\gamma)$ in Eq. (2.6). This metric is sometimes called canonical 5D space (C_5), and in many respects takes over the role of 5D Minkowski space (M_5). It may be viewed as a kind of spherical embedding for 4D spacetime, where $x^4 = \ell$ acts like a radius. When the conditions for pure-canonical space are relaxed, the cosmological 'constant' becomes variable. In particular, when ℓ is shifted by a constant in Eq. (2.6), a singular hypersurface appears similar to that of M theory, and Λ diverges there, explaining the 4D big bang.

The preceding paragraphs, while reviewing the main points of M theory and STM theory, do not take a position on the important question of whether the fifth dimension is merely a mathematical artifice or is 'real' in the usual sense of the word. We propose, in view of the controversial nature of this question, to defer a discussion of it to Chapter 9. In the meantime, we will examine the implications of STM theory in a systematic manner, using $x^4 = \ell$ as just another coordinate. We will begin, in the next section, with an important proof that an empty manifold in 5D can lead to matter in 4D. This is in the nature of a mathematical exercise, and we will return to physical models in Section 2.4.

2.3. Field Equations for Neutral Matter

In this section, we will take the 5D field equations and break them down to 4D relations, for the broad class of physical problems where there are gravitational and scalar fields, but no electromagnetic field. We start with $R_{AB} = 0$ $(A, B = 0 - 4)$ and end with Einstein's field equations $R_{\alpha\beta} = 8\pi T_{\alpha\beta}$ $(\alpha, \beta = 0 - 3)$ plus other relations.

Any fully covariant 5D theory, such as STM theory, has five coordinate degrees of freedom, which used judiciously can lead to considerable algebraic simplification without loss of generality. Therefore, a natural case to study is specified by $g_{4\alpha} = 0$, $g_{44} \neq 0$. This removes the explicit electromagnetic potentials and leaves one coordinate degree of freedom over to be used appropriately (e.g., to simplify the equations of motion of a particle). This choice of coordinates or 'gauge choice' involves $g_{\alpha\beta} = g_{\alpha\beta}(x^A)$, $g_{44} = g_{44}(x^A)$ and so is not restricted by the cylinder condition of old Kaluza-Klein theory. Neither do we impose compactification. The working below follows that of Wesson and Ponce de Leon from 1992 [8], and has many applications.

We take a 5D interval $dS^2 = g_{AB} dx^A dx^B$ where

$$g_{\alpha\beta} = g_{\alpha\beta}(x^A) \quad , \quad g_{4\alpha} = 0 \quad ,$$
$$g_{44} \equiv \varepsilon\Phi^2(x^A) \quad , \quad g^{44} = \frac{1}{g_{44}} = \frac{\varepsilon}{\Phi^2} \quad . \tag{2.7}$$

Here $\varepsilon^2 = 1$ and the signature of the scalar part of the metric is left general. (We will see later that there are well-behaved classical solutions of the field equations with $\varepsilon = +1$ as well as the often-assumed $\varepsilon = -1$, and the freedom to choose this may also help with the Euclidean approach to quantum gravity.) The 5D Ricci tensor in terms of the 5D Christoffel symbols is given by

$$R_{AB} = \partial_C \Gamma^C_{AB} - \partial_B \Gamma^C_{AC} + \Gamma^C_{AB}\Gamma^D_{CD} - \Gamma^C_{AD}\Gamma^D_{BC} \quad . \tag{2.8}$$

Here ∂_A (and below, ∇_A) denote ordinary (and covariant) derivatives respectively, but now taken with respect to all five coordinates x^A. Putting $A \to \alpha, B \to \beta$ in Eq. (2.8) gives us the 4D part of the 5D quantity. For convenience, we will also use an overstar (*) to denote the partial derivative with respect to $x^4 = \ell$. Expanding some summed terms on the right-hand side by letting $C \to \lambda, 4$ etc. and rearranging gives

$$\hat{R}_{\alpha\beta} = \partial_\lambda\Gamma^\lambda_{\alpha\beta} + \partial_4\Gamma^4_{\alpha\beta} - \partial_\beta\Gamma^\lambda_{\alpha\lambda} - \partial_\beta\Gamma^4_{\alpha 4} + \Gamma^\lambda_{\alpha\beta}\Gamma^\mu_{\lambda\mu}$$
$$+ \Gamma^\lambda_{\alpha\beta}\Gamma^4_{\lambda 4} + \Gamma^4_{\alpha\beta}\Gamma^D_{4D} - \Gamma^\mu_{\alpha\lambda}\Gamma^\lambda_{\beta\mu} - \Gamma^4_{\alpha\lambda}\Gamma^\lambda_{\beta 4} - \Gamma^D_{\alpha 4}\Gamma^4_{\beta D} \quad . \tag{2.9}$$

Here and elsewhere, a hat denotes the 4D part of a 5D quantity. Part of this is the conventional Ricci tensor $R_{\alpha\beta}$ (that only depends on indices 0-3), so

$$\hat{R}_{\alpha\beta} = R_{\alpha\beta} + \partial_4\Gamma^4_{\alpha\beta} - \partial_\beta\Gamma^4_{\alpha 4} + \Gamma^\lambda_{\alpha\beta}\Gamma^4_{\lambda 4}$$
$$+ \Gamma^4_{\alpha\beta}\Gamma^D_{4D} - \Gamma^4_{\alpha\lambda}\Gamma^\lambda_{\beta 4} - \Gamma^D_{\alpha 4}\Gamma^4_{\beta D} \quad . \tag{2.10}$$

To evaluate this we need the Christoffel symbols.

It is convenient to tabulate these (with some redundancy) in three separate groups, the first of which reads:

$$\Gamma^4_{\alpha\beta} = -\tfrac{1}{2}g^{44}\overset{*}{g}_{\alpha\beta} \quad , \quad \Gamma^4_{\alpha 4} = \tfrac{1}{2}g^{44}\partial_\alpha g_{44} \quad ,$$

$$\Gamma^D_{4D} = \tfrac{1}{2} g^{DC} \overset{*}{g}_{DC} \quad , \quad \Gamma^{\lambda}_{\beta 4} = \tfrac{1}{2} g^{\lambda C} \overset{*}{g}_{\beta C} \quad ,$$

$$\Gamma^D_{\alpha 4} = \tfrac{1}{2} \left(g^{D4} \partial_{\alpha} g_{44} + g^{D\gamma} \overset{*}{g}_{\alpha\gamma} \right) \quad , \tag{2.11}$$

$$\Gamma^4_{\beta D} = \tfrac{1}{2} \left(g^{44} \partial_{\beta} g_{D4} - g^{44} \overset{*}{g}_{\beta D} \right) \quad .$$

The second group reads:

$$\Gamma^{\lambda}_{44} = -\tfrac{1}{2} g^{\lambda\beta} \partial_{\beta} g_{44} \quad , \quad \Gamma^{\lambda}_{4\lambda} = \tfrac{1}{2} g^{\lambda\beta} \overset{*}{g}_{\lambda\beta} \quad ,$$

$$\Gamma^{\mu}_{\lambda\mu} = \tfrac{1}{2} g^{\mu\beta} \partial_{\lambda} g_{\mu\beta} \quad , \quad \Gamma^4_{44} = \tfrac{1}{2} g^{44} \overset{*}{g}_{44} \quad , \tag{2.12}$$

$$\Gamma^{\lambda}_{4\mu} = \tfrac{1}{2} g^{\lambda\beta} \overset{*}{g}_{\mu\beta} \quad , \quad \Gamma^4_{4\mu} = \tfrac{1}{2} g^{44} \partial_{\mu} g_{44} \quad .$$

The third group reads:

$$\Gamma^{\lambda}_{4\alpha} = \tfrac{1}{2} g^{\lambda\mu} \overset{*}{g}_{\alpha\mu} \quad , \quad \Gamma^4_{4\alpha} = \tfrac{1}{2} g^{44} \partial_{\alpha} g_{44} \quad ,$$

$$\Gamma^{\lambda}_{4\lambda} = \tfrac{1}{2} g^{\lambda\beta} \overset{*}{g}_{\lambda\beta} \quad , \quad \Gamma^{\mu}_{\lambda\mu} = \tfrac{1}{2} g^{\mu\nu} \partial_{\lambda} g_{\mu\nu} \quad ,$$

$$\Gamma^{\mu}_{\alpha\lambda} = \tfrac{1}{2} g^{\mu\sigma} (\partial_{\alpha} g_{\lambda\sigma} + \partial_{\lambda} g_{\sigma\alpha} - \partial_{\sigma} g_{\alpha\lambda}) \quad , \tag{2.13}$$

$$\Gamma^{\lambda}_{44} = -\tfrac{1}{2} g^{\alpha\lambda} \partial_{\alpha} g_{44} \quad , \quad \Gamma^4_{\alpha\lambda} = -\tfrac{1}{2} g^{44} \overset{*}{g}_{\alpha\lambda} \quad .$$

We will use the expressions in Eqs. (2.11), (2.12) and (2.13) respectively to evaluate Eq. (2.10) above and Eqs. (2.17) and (2.27) below.

Thus substituting into and expanding some terms in Eq. (2.10) gives

$$\hat{R}_{\alpha\beta} = R_{\alpha\beta} - \tfrac{1}{2} \overset{*}{g}^{44} \overset{*}{g}_{\alpha\beta} - \tfrac{1}{2} g^{44} \overset{**}{g}_{\alpha\beta} - \tfrac{1}{2} \partial_{\beta} g^{44} \partial_{\alpha} g_{44}$$

$$-\frac{1}{2}g^{44}\partial_\alpha\partial_\beta g_{44} + \frac{1}{2}g^{44}\partial_\lambda g_{44}\Gamma^\lambda_{\alpha\beta} - \frac{1}{4}g^{\mu\nu}\overset{*}{g}_{\mu\nu}g^{44}\overset{*}{g}_{\alpha\beta}$$

$$-\frac{1}{4}(g^{44})^2\overset{*}{g}_{\alpha\beta}\overset{*}{g}_{44} + \frac{1}{2}g^{\lambda\mu}g^{44}\overset{*}{g}_{\alpha\lambda}\overset{*}{g}_{\beta\mu}$$

$$-\frac{1}{4}(g^{44})^2\partial_\alpha g_{44}\partial_\beta g_{44} \quad . \tag{2.14}$$

Some of the terms here may be rewritten using

$$-\frac{1}{2}\partial_\beta g^{44}\partial_\alpha g_{44} - \frac{1}{2}g^{44}\partial_\alpha\partial_\beta g_{44} + \frac{1}{2}g^{44}\partial_\lambda g_{44}\Gamma^\lambda_{\alpha\beta}$$

$$-\frac{1}{4}(g^{44})^2\partial_\alpha g_{44}\partial_\beta g_{44} = -\frac{1}{\Phi}\left(\partial_\beta\Phi_\alpha - \Gamma^\lambda_{\alpha\beta}\Phi_\lambda\right)$$

$$\equiv -\frac{\nabla_\beta\Phi_\alpha}{\Phi} \quad , \tag{2.15}$$

where $\Phi_\alpha \equiv \partial_\alpha\Phi$. Then Eq. (2.14) gives

$$\hat{R}_{\alpha\beta} = R_{\alpha\beta} - \frac{\nabla_\beta\Phi_\alpha}{\Phi}$$

$$+ \frac{\varepsilon}{2\Phi^2}\left(\frac{\Phi\overset{*}{g}_{\alpha\beta}}{\Phi} - \overset{**}{g}_{\alpha\beta} + g^{\lambda\mu}\overset{*}{g}_{\alpha\lambda}\overset{*}{g}_{\beta\mu} - \frac{1}{2}g^{\mu\nu}\overset{*}{g}_{\mu\nu}\overset{*}{g}_{\alpha\beta}\right) \quad . \tag{2.16}$$

We will use this below when we consider the field equations.

Returning to Eq. (2.8), we put the indices $A = 4$, $B = 4$ and expand with $C \to \lambda, 4$ etc. to obtain

$$R_{44} = \partial_\lambda\Gamma^\lambda_{44} - \partial_4\Gamma^\lambda_{4\lambda} + \Gamma^\lambda_{44}\Gamma^\mu_{\lambda\mu} + \Gamma^4_{44}\Gamma^\mu_{4\mu} - \Gamma^\lambda_{4\mu}\Gamma^\mu_{4\lambda} - \Gamma^4_{4\mu}\Gamma^\mu_{44} \quad . \tag{2.17}$$

The Christoffel symbols here are tabulated in Eqs. (2.12), and cause Eq. (2.17) to become

$$R_{44} = -\frac{1}{2}\partial_\lambda g^{\lambda\beta}\partial_\beta g_{44} - \frac{1}{2}g^{\lambda\beta}\partial_\lambda\partial_\beta g_{44} - \frac{1}{2}\overset{*}{g}^{\lambda\beta}\overset{*}{g}_{\lambda\beta} - \frac{1}{2}g^{\lambda\beta}\overset{**}{g}_{\lambda\beta}$$

$$-\frac{1}{4}g^{\lambda\beta}\partial_\beta g_{44}g^{\mu\sigma}\partial_\lambda g_{\mu\sigma} + \frac{1}{4}g^{44}\overset{*}{g}_{44}g^{\lambda\beta}\overset{*}{g}_{\lambda\beta}$$

$$-\frac{1}{4}g^{\mu\beta}\overset{*}{g}_{\lambda\beta}g^{\lambda\sigma}\overset{*}{g}_{\mu\sigma} + \frac{1}{4}g^{44}\partial_\lambda g_{44}g^{\lambda\beta}\partial_\beta g_{44} \quad . \tag{2.18}$$

Some of the terms here may be rewritten using

$$-\frac{1}{2}\partial_\lambda g^{\lambda\beta}\partial_\beta g_{44} - \frac{1}{2}g^{\lambda\beta}\partial_\lambda\partial_\beta g_{44} - \frac{1}{4}g^{\lambda\beta}\partial_\beta g_{44}g^{\mu\sigma}\partial_\lambda g_{\mu\sigma}$$
$$+\frac{1}{4}g^{44}\partial_\lambda g_{44}g^{\lambda\beta}\partial_\beta g_{44}$$
$$= -\varepsilon\Phi\left(\partial_\lambda g^{\lambda\beta}\Phi_\beta + g^{\lambda\beta}\partial_\lambda\Phi_\beta + \frac{1}{2}g^{\lambda\beta}g^{\mu\sigma}\partial_\lambda g_{\mu\sigma}\Phi_\beta\right)$$
$$= -\varepsilon\Phi g^{\mu\nu}\nabla_\nu\Phi_\mu \quad . \tag{2.19}$$

Here we have obtained the last line by noting that $\nabla_\lambda\Phi_\beta = \partial_\lambda\Phi_\beta - \Gamma^\sigma_{\beta\lambda}\Phi_\sigma$ implies

$$g^{\beta\lambda}\partial_\lambda\Phi_\beta + \frac{1}{2}g^{\beta\lambda}g^{\sigma\mu}\partial_\mu g_{\beta\lambda}\,\Phi_\sigma = g^{\mu\nu}\nabla_\nu\Phi_\mu$$
$$+\frac{1}{2}g^{\beta\lambda}g^{\sigma\mu}\partial_\lambda g_{\mu\beta}\,\Phi_\sigma + \frac{1}{2}g^{\beta\lambda}g^{\sigma\mu}\partial_\beta g_{\mu\lambda}\,\Phi_\sigma \quad , \tag{2.20}$$

and that $\partial_\mu\delta^\mu_\nu = 0$ implies $\left(\partial_\lambda g^{\lambda\sigma} + g^{\beta\lambda}g^{\sigma\mu}\partial_\lambda g_{\mu\beta}\right)\Phi_\sigma = 0$. Putting Eq. (2.19) in Eq. (2.18) gives lastly

$$R_{44} = -\varepsilon\Phi\Box\Phi - \frac{1}{2}\overset{*}{g}^{\lambda\beta}\overset{*}{g}_{\lambda\beta} - \frac{1}{2}g^{\lambda\beta}\overset{**}{g}_{\lambda\beta}$$
$$+\frac{\overset{*}{\Phi}}{2\Phi}g^{\lambda\beta}\overset{*}{g}_{\lambda\beta} - \frac{1}{4}g^{\mu\beta}g^{\lambda\sigma}\overset{*}{g}_{\lambda\beta}\overset{*}{g}_{\mu\sigma} \quad , \tag{2.21}$$

where $\Box\Phi \equiv g^{\mu\nu}\nabla_\nu\Phi_\mu$ defines the 4D curved-space box operator (D'Alembertian). Equations (2.16) and (2.21) can be used with the 5D field equations, which we repeat here:

$$R_{AB} = 0 \quad . \tag{2.22}$$

Then $\hat{R}_{\alpha\beta} = 0$ in Eq. (2.16) gives

$$R_{\alpha\beta} = \frac{\nabla_\beta\Phi_\alpha}{\Phi} - \frac{\varepsilon}{2\Phi^2}\left(\frac{\overset{*}{\Phi}\overset{*}{g}_{\alpha\beta}}{\Phi} - \overset{**}{g}_{\alpha\beta} + g^{\lambda\mu}\overset{*}{g}_{\alpha\lambda}\overset{*}{g}_{\beta\mu} - \right.$$
$$\left. \frac{1}{2}g^{\mu\nu}\overset{*}{g}_{\mu\nu}\overset{*}{g}_{\alpha\beta}\right) \quad . \tag{2.23}$$

And $R_{44} = 0$ in Eq. (2.21) gives

$$\varepsilon\Phi\Box\Phi = -\frac{1}{4}\overset{*}{g}^{\lambda\beta}\overset{*}{g}_{\lambda\beta} - \frac{1}{2}g^{\lambda\beta}\overset{**}{g}_{\lambda\beta} + \left(\frac{\overset{*}{\Phi}}{2\Phi}\right)g^{\lambda\beta}\overset{*}{g}_{\lambda\beta} \quad , \qquad (2.24)$$

where we have noted that $\partial_4\delta_\nu^\mu = 0$ implies that $g^{\mu\beta}g^{\lambda\sigma}\overset{*}{g}_{\lambda\beta}\overset{*}{g}_{\mu\sigma} + \overset{*}{g}^{\mu\sigma}\overset{*}{g}_{\mu\sigma} = 0$. From Eq. (2.23) we can form the 4D Ricci curvature scalar $R = g^{\alpha\beta}R_{\alpha\beta}$. Eliminating the covariant derivative using Eq. (2.24), and again using $\partial_4\delta_\nu^\mu = 0$ to eliminate some terms, gives

$$R = \frac{\varepsilon}{4\Phi^2}\left[\overset{*}{g}^{\mu\nu}\overset{*}{g}_{\mu\nu} + \left(g^{\mu\nu}\overset{*}{g}_{\mu\nu}\right)^2\right] \quad . \qquad (2.25)$$

With Eqs. (2.23) and (2.25), we are now in a position to define if we wish an energy-momentum tensor in 4D via $8\pi T_{\alpha\beta} \equiv R_{\alpha\beta} - Rg_{\alpha\beta}/2$. It is

$$8\pi T_{\alpha\beta} = \frac{\nabla_\beta\Phi_\alpha}{\Phi} - \frac{\varepsilon}{2\Phi^2}\left\{\frac{\overset{*}{\Phi}\overset{*}{g}_{\alpha\beta}}{\Phi} - \overset{**}{g}_{\alpha\beta} + g^{\lambda\mu}\overset{*}{g}_{\alpha\lambda}\overset{*}{g}_{\beta\mu} - \right.$$
$$\left. -\frac{1}{2}g^{\mu\nu}\overset{*}{g}_{\mu\nu}\overset{*}{g}_{\alpha\beta} + \frac{1}{4}g_{\alpha\beta}\left[\overset{*}{g}^{\mu\nu}\overset{*}{g}_{\mu\nu} + \left(g^{\mu\nu}\overset{*}{g}_{\mu\nu}\right)^2\right]\right\} \quad . \qquad (2.26)$$

Provided we use this energy-momentum tensor, Einstein's 4D field equations $G_{\alpha\beta} = 8\pi T_{\alpha\beta}$ will be satisfied *automatically*. Matter and energy have thus been 'induced' in 4D from pure geometry in 5D. For this reason, STM theory is sometimes also referred to as 'Induced Matter Theory,' and $T_{\alpha\beta}$ as defined by Eq. (2.26) is known as the induced-matter energy-momentum tensor.

The mathematical expression (2.26) has good properties. It is a symmetric tensor that has a part which depends on derivatives of Φ with respect to the usual coordinates x^{0123}, and a part which depends on derivatives of other metric coefficients with respect to the extra coordinate x^4. [The first term in Eq. (2.26) is implicitly symmetric because it depends on the second partial derivative, while the other terms are explicitly symmetric.] It is also compatible with what is known about the recovery of 4D properties of matter from apparently empty 5D solutions of Kaluza-Klein theory. With Eq. (2.26) and preceding relations, the case where there is no dependency on x^4 becomes transparent. Then Eq. (2.24) becomes the scalar wave equation for the extra part of the metric ($g^{\mu\nu}\nabla_\nu\Phi_\mu = 0$ with $g_{44} = \varepsilon\Phi^2$). And Eq. (2.26) gives $T = T_{\alpha\beta}g^{\alpha\beta}$, which implies a radiation-like equation of state. However, in general there must be x^4-dependence if we are to recover more complex equations of state from solutions of $R_{AB} = 0$.

These field equations have four other components we have not so far considered, namely $R_{4\alpha} = 0$. This relation by Eq. (2.8) expanded is

$$R_{4\alpha} = \partial_\lambda\Gamma_{4\alpha}^\lambda + \partial_4\Gamma_{4\alpha}^4 - \partial_\alpha\Gamma_{4\alpha}^\lambda - \partial_\alpha\Gamma_{44}^4 + \Gamma_{4\alpha}^\lambda\Gamma_{\lambda A}^A$$
$$+\Gamma_{4\alpha}^4\Gamma_{4A}^A - \Gamma_{4\mu}^A\Gamma_{\alpha A}^\mu - \Gamma_{44}^D\Gamma_{\alpha D}^4 \quad . \tag{2.27}$$

The Christoffel symbols here are tabulated in Eq. (2.13) and cause Eq. (2.27) to become

$$R_{4\alpha} = \frac{1}{4}g^{44}g^{\lambda\beta}\left(\overset{*}{g}_{\lambda\beta}\partial_\alpha g_{44} - \partial_\beta g_{44}\overset{*}{g}_{\alpha\lambda}\right) + \frac{1}{2}\partial_\lambda g^{\lambda\mu}\overset{*}{g}_{\mu\alpha} +$$
$$\frac{1}{2}g^{\lambda\mu}\partial_\lambda\overset{*}{g}_{\mu\alpha} - \frac{1}{2}\partial_\alpha g^{\lambda\beta}\overset{*}{g}_{\lambda\beta} - \frac{1}{2}g^{\lambda\beta}\partial_\alpha\overset{*}{g}_{\lambda\beta} +$$
$$\frac{1}{4}g^{\lambda\sigma}g^{\mu\beta}\overset{*}{g}_{\sigma\alpha}\partial_\lambda g_{\mu\beta} + \frac{1}{4}\overset{*}{g}^{\mu\beta}\partial_\alpha g_{\mu\beta} \quad . \tag{2.28}$$

Here we have done some algebra using $\partial_{(\alpha \text{ and } 4)}(g_{44}g^{44}) = 0$ or $\overset{*}{g}^{44}\partial_\alpha g_{44} - \partial_\alpha g^{44}\overset{*}{g}_{44} = 0$, and $\partial_4\delta_\lambda^\beta = 0$ or $g^{\lambda\nu}g_{\mu\nu}g^{\mu\sigma} + \overset{*}{g}^{\lambda\sigma} = 0$. [We also note in passing that one can use $\partial_4\Gamma_{4\alpha}^4 = \partial_\alpha\Gamma_{44}^4$ in Eq. (2.27) and obtain an alternative form of Eq. (2.28) with the last term replaced by $\partial_\alpha g^{\mu\beta}\overset{*}{g}_{\mu\beta}/4$.] While Eq. (2.28) may be useful in other computations, it is helpful for our purpose here to rewrite it as

$$R_{4\alpha} = \frac{1}{2}\partial_\beta\left(g^{\beta\lambda}\overset{*}{g}_{\lambda\alpha}\right) - \frac{1}{2}\partial_\alpha\left(g^{\mu\nu}\overset{*}{g}_{\mu\nu}\right) +$$
$$\frac{1}{4}g^{\mu\beta}\partial_\lambda g_{\mu\beta}g^{\lambda\sigma}\overset{*}{g}_{\sigma\alpha} - \frac{1}{4}g^{\lambda\beta}\partial_\alpha g_{\beta\mu}g^{\mu\sigma}\overset{*}{g}_{\sigma\lambda} -$$
$$\frac{1}{4}g^{44}\partial_\beta g_{44}\left(g^{\beta\lambda}\overset{*}{g}_{\lambda\alpha} - \delta_\alpha^\beta g^{\mu\nu}\overset{*}{g}_{\mu\nu}\right) \quad . \tag{2.29}$$

Noting that $\partial/\partial x^\alpha = \delta_\alpha^\beta (\partial/\partial x^\beta)$ and moreover that $-g^{44}\partial_\beta g_{44}/2 = \sqrt{g_{44}} (\partial/\partial x^\beta)(1/\sqrt{g_{44}})$ allows us to obtain finally

$$\frac{R_{4\alpha}}{\sqrt{g_{44}}} = \frac{\partial}{\partial x^\beta}\left[\frac{1}{2\sqrt{g_{44}}}\left(g^{\beta\lambda}\overset{*}{g}_{\lambda\alpha} - \delta_\alpha^\beta g^{\mu\nu}\overset{*}{g}_{\mu\nu}\right)\right]$$
$$+ \frac{1}{4\sqrt{g_{44}}}\left(g^{\mu\beta}\partial_\lambda g_{\mu\beta}g^{\lambda\sigma}\overset{*}{g}_{\sigma\alpha} - g^{\lambda\beta}\partial_\alpha g_{\beta\mu}g^{\mu\sigma}\overset{*}{g}_{\sigma\lambda}\right) \ . \quad (2.30)$$

This form suggests we should introduce the 4-tensor

$$P_\alpha^\beta \equiv \frac{1}{2\sqrt{g_{44}}}\left(g^{\beta\lambda}\overset{*}{g}_{\lambda\alpha} - \delta_\alpha^\beta g^{\mu\nu}\overset{*}{g}_{\mu\nu}\right) \ . \quad (2.31)$$

The divergence of this is

$$\nabla_\beta P_\alpha^\beta = \partial_\beta P_\alpha^\beta + \Gamma_{\beta\mu}^\beta P_\alpha^\mu - \Gamma_{\alpha\beta}^\mu P_\mu^\beta \quad ,$$

which when written out in full may be shown to be the same as the right-hand side of Eq. (2.30). The latter therefore reads

$$\frac{R_{4\alpha}}{\sqrt{g_{44}}} = \nabla_\beta P_\alpha^\beta \ . \quad (2.32)$$

The subset of the field equations (2.22) given by $R_{4\alpha} = 0$ can then be summed up by the relations

$$\nabla_\beta P_\alpha^\beta = 0 \quad ,$$
$$P_\alpha^\beta \equiv \frac{1}{2\sqrt{g_{44}}}\left(g^{\beta\sigma}\overset{*}{g}_{\sigma\alpha} - \delta_\alpha^\beta g^{\mu\nu}\overset{*}{g}_{\mu\nu}\right) \ . \quad (2.33)$$

These have the appearance of conservation laws for P_α^β, sometimes referred to in STM theory as the 'P-tensor'. Its fully covariant form and associated scalar are:

$$P_{\alpha\beta} = \frac{1}{2\sqrt{g_{44}}}\left(\overset{*}{g}_{\alpha\beta} - g_{\alpha\beta}g^{\mu\nu}\overset{*}{g}_{\mu\nu}\right) \quad ,$$

$$\quad (2.34)$$

$$P = \frac{-3g^{\lambda\sigma}\overset{*}{g}_{\lambda\sigma}}{2\sqrt{g_{44}}} \quad .$$

We will examine these quantities elsewhere. It should be noted that P_α^β and its associated quantities differ from others in having physical dimensions (units) of an inverse length, so it may have a new physical meaning.

It is apparent from the working in this section that the starting conditions (2.7) provide a convenient way to split the 5D field equations $R_{AB} = 0$ into three sets: The 5D equations $\hat{R}_{\alpha\beta} = 0$ give a set of equations in the 4D Ricci tensor $R_{\alpha\beta}$, Eq. (2.23); the 5D equation $R_{44} = 0$ gives a wave-like equation in the scalar potential, Eq. (2.24); and the 5D equations $R_{4\alpha} = 0$ can be expressed as a set of 4D conservation laws, Eqs. (2.33). Along the way we also obtain some other useful relations, notably an expression for the 4D Ricci scalar in terms of the dependence of the 4D metric on the extra coordinate, Eq. (2.25). However, the physically most relevant expression is an effective or induced 4D energy-momentum tensor, defined by Eq. (2.26). Another way to express these results is to say that the 15 field equations $R_{AB} = 0$ or $G_{AB} = 0$ can always be split into three sets which make physical sense provided the metric is allowed to depend on the extra coordinate x^4. These sets consist of four conservation equations, one wave equation for the scalar field, and ten equations for fields and matter of gravitational type. In fact, the last *are* Einstein's equations of general relativity, with matter induced from the extra dimension.

2.4. Some Simple Solutions

We wish to quote some illustrative solutions of the field equations $R_{AB} = 0$ considered in the previous section. In practice, these equations are best approached by solving Eqs. (2.24) and (2.33) first, and then balancing the Einstein equations (2.26) and so obtaining the source. However, satisfying the field equations presupposes a form for the metric, which in turn presumes some information about the physics involved. The solutions quoted below therefore anticipate detailed work done later; notably, on the standard cosmological models in Section 3.2 and the one-body problem in Section 7.3. The solutions for these cases will serve to show that the STM approach gives reasonable physical results. In addition, we wish to show in a more step-by-step manner exactly how it is that the field equations are separated into parts for the fields and their sources. To do this, we will revisit one of the three calculations made in the original work on STM, namely that concerning cosmological models with flat spatial sections [9]. (The other two calculations concerned the energy-momentum tensor of the preceding section and the solitons discussed in Chapter 5, and were first studied in Refs. [8] and [10]). By reconsidering this problem, it will become clear how an apparently empty 5D space can be decomposed to yield a 4D space and matter.

To obtain 5D analogs of all the 4D Robertson-Walker models with flat 3D sections, it is necessary to choose an appropriate metric. Such is

$$dS^2 = e^\nu dt^2 - e^\omega (dr^2 + r^2 d\Omega^2) - e^\mu d\ell^2 \quad . \quad (2.35)$$

Here the time coordinate $x^0 = t$ and the space coordinates $x^{123} = r\theta\phi$ ($d\Omega^2 \equiv d\theta^2 + \sin^2\theta \, d\phi^2$) have been augmented by the new coordinate $x^4 = \ell$. The metric coefficients ν, ω and μ will depend in general on both t and ℓ, partial derivatives with respect to which will be denoted by an overdot and an overstar respectively. The components of the Einstein tensor in mixed form are:

$$G_0^0 = e^{-\nu}\left(-\frac{3\dot\omega^2}{4} - \frac{3\dot\omega\dot\mu}{4}\right) + e^{-\mu}\left(\frac{3\overset{**}{\omega}}{2} + \frac{3\overset{*}{\omega}^2}{2} - \frac{3\overset{*}{\mu}\overset{*}{\omega}}{4}\right) \, ,$$

$$G_4^0 = e^{-\nu}\left(\frac{3\overset{*}{\dot\omega}}{2} + \frac{3\overset{*}{\omega}\dot\omega}{4} - \frac{3\overset{*}{\omega}\dot\nu}{4} - \frac{3\dot\omega\overset{*}{\mu}}{4}\right) \, ,$$

$$G_1^1 = G_2^2 = G_3^3 = -e^{-\nu}\left(\ddot\omega + \frac{3\dot\omega^2}{4} + \frac{\ddot\mu}{2} + \frac{\dot\mu^2}{4} + \frac{\dot\omega\dot\mu}{2} - \frac{\dot\nu\dot\omega}{2} - \frac{\dot\nu\dot\mu}{4}\right) +$$
$$e^{-\mu}\left(\overset{**}{\omega} + \frac{3\overset{*}{\omega}^2}{4} + \frac{\overset{**}{\nu}}{2} + \frac{\overset{*}{\nu}^2}{4} + \frac{\overset{*}{\omega}\overset{*}{\nu}}{2} - \frac{\overset{*}{\mu}\overset{*}{\omega}}{2} - \frac{\overset{*}{\nu}\overset{*}{\mu}}{4}\right) \, ,$$

$$G_4^4 = -e^{-\nu}\left(\frac{3\ddot\omega}{2} + \frac{3\dot\omega^2}{2} - \frac{3\dot\nu\dot\omega}{4}\right) + e^{-\mu}\left(\frac{3\overset{*}{\omega}^2}{4} + \frac{3\overset{*}{\omega}\overset{*}{\nu}}{4}\right) \, . \quad (2.36)$$

These are 5D components. We wish to match the terms here with the components of the usual 4D perfect-fluid energy-momentum tensor. This is $T_{\alpha\beta} = (p + \rho)u_\alpha u_\beta - pg_{\alpha\beta}$, where $u^\alpha \equiv dx^\alpha/ds$, and for our case has components $T_0^0 = \rho$, $T_1^1 = -p$ for the density and pressure. Following the philosophy of STM, we simply identify the new terms (due to the fifth dimension) in G_0^0 with ρ, and the new terms in G_1^1 with p. Then, collecting terms which depend on the new metric coefficient μ or derivatives with respect to the new coordinate ℓ, we define

$$8\pi\rho \equiv -\frac{3}{4}e^{-\nu}\dot{\omega}\dot{\mu} + \frac{3}{2}e^{-\mu}\left(\overset{**}{\omega} + \overset{*}{\omega}{}^2 - \frac{\overset{*}{\mu}\overset{*}{\omega}}{2}\right) \ ,$$

$$8\pi p \equiv e^{-\nu}\left(\frac{\ddot{\mu}}{2} + \frac{\dot{\mu}^2}{4} + \frac{\dot{\omega}\dot{\mu}}{2} - \frac{\dot{\nu}\dot{\mu}}{4}\right)$$
$$-e^{-\mu}\left(\overset{**}{\omega} + \frac{3\overset{*}{\omega}{}^2}{4} + \frac{\overset{**}{\nu}}{2} + \frac{\overset{*}{\nu}{}^2}{4} + \frac{\overset{*}{\omega}\overset{*}{\nu}}{2} - \frac{\overset{*}{\mu}\overset{*}{\omega}}{2} - \frac{\overset{*}{\nu}\overset{*}{\mu}}{4}\right) \ .$$
(2.37)

These are suggested identifications for the 4D properties of matter in terms of 5D properties of geometry.

To see if they make physical sense to this point, we combine Eqs. (2.37) and (2.36) with the field equations $G_B^A = 0$. Then

$$G_0^0 = -\frac{3}{4}e^{-\nu}\dot{\omega}^2 + 8\pi\rho = 0 \ ,$$

$$G_4^0 = e^{-\nu}\left(\frac{3}{2}\overset{*}{\dot{\omega}} + \frac{3}{4}\dot{\omega}\overset{*}{\omega} - \frac{3}{4}\dot{\omega}\overset{*}{\nu} - \frac{3}{4}\overset{*}{\omega}\dot{\mu}\right) = 0 \ ,$$

$$G_1^1 = -e^{-\nu}\left(\ddot{\omega} + \frac{3}{4}\dot{\omega}^2 - \frac{\dot{\nu}\dot{\omega}}{2}\right) - 8\pi p = 0 \ ,$$
(2.38)

$$G_4^4 = -e^{-\nu}\left(\frac{3}{2}\ddot{\omega} + \frac{3\dot{\omega}^2}{2} - \frac{3\dot{\nu}\dot{\omega}}{4}\right) + e^{-\mu}\left(\frac{3}{4}\overset{*}{\omega}{}^2 + \frac{3\overset{*}{\omega}\overset{*}{\nu}}{4}\right) = 0 \ .$$

We see from the first of these equations that ρ must be positive; and from the third that p could in principle be negative, as needed in classical descriptions of particle production in quantum field theory. To make further progress, however, we need solutions of the field equations $R_{AB} = 0$ or $G_{AB} = 0$ which are more explicit.

A simple solution which is well known but does not depend on ℓ has $\nu = 0$, $\omega = \log t$ and $\mu = -\ln t$ in Eq. (2.35), which now reads

$$dS^2 = dt^2 - t(dr^2 + r^2 d\Omega^2) - t^{-1}d\ell^2 \ .$$
(2.39)

This has a shrinking fifth dimension, and from Eqs. (2.37) or (2.38) density and pressure given by $8\pi\rho = 3/4t^2$ and $8\pi p = 1/4t^2$. If these are combined to form the gravitational density $(\rho + 3p)$ and the proper radial distance $R \equiv e^{\omega/2}r$ is introduced, then the mass of a portion of the fluid is $M = 4\pi R^3(\rho + 3p)/3$. The field equations then lead to the usual law of motion $\ddot{R} = -M/R^2$. Similarly, the first law of thermodynamics is recovered by writing $dE + pdV = 0$ as $(\rho e^{3\omega/2}) + p(e^{3\omega/2}) = 0$, where E is the energy and V is the 3D volume. The equation of state of the fluid described by Eq. (2.39) is the $p = \rho/3$ typical of radiation.

To go beyond radiation, we use one of the solutions of $R_{AB} = 0$ due to Ponce de Leon [18]. With a redefinition of constants appropriate to STM theory, it has $e^{\nu} = \ell^2$, $e^{\omega} = t^{2/\alpha}\ell^{2/(1-\alpha)}$, $e^{\mu} = \alpha^2(1 - \alpha)^{-2}t^2$ in Eq. (2.35), which now reads

$$dS^2 = \ell^2 dt^2 - t^{2/\alpha}\ell^{2/(1-\alpha)}(dr^2 + r^2 d\Omega^2)$$
$$-\alpha^2(1 - \alpha)^{-2}t^2 d\ell^2 \quad . \tag{2.40}$$

This has a growing fifth dimension, and density and pressure which depend on the one assignable constant α. From Eqs. (2.37) or (2.38) these are given by

$$8\pi\rho = \frac{3}{\alpha^2\ell^2 t^2} \quad , \quad 8\pi p = \frac{2\alpha-3}{\alpha^2\ell^2 t^2} \quad . \tag{2.41}$$

The presence of ℓ here may appear puzzling at first, but the coordinates are of course arbitrary and the proper time is $T \equiv \ell t$. (Alternatively, the presence of $x^4 = \ell$ depends on whether we consider the pure 4D metric or the 4D part of the 5D metric.) In terms of this, $8\pi\rho = 3/\alpha^2 T^2$ and $8\pi p = (2\alpha - 3)/\alpha^2 T^2$. For $\alpha = 3/2$, $8\pi\rho = 4/3T^2$ and $p = 0$. While for $\alpha = 2$, $8\pi\rho = 3/4T^2$ and $8\pi p = 1/4T^2$. The former is identical to the 4D Einstein-de Sitter model for the late universe with dust. The latter is identical to the 4D standard model for the early universe with radiation or highly relativistic particles. (The coincidence of the properties of matter for this model with $\alpha = 2$ and the previous model does not necessarily imply that they are the same, since similar matter can belong to different solutions even in 4D.) As before, the usual forms of the law of motion and the first law of thermodynamics are recovered.

Other properties of the class of solutions with metric (2.40) are discussed in Section 3.2, where a coordinate transformation is given which shows that the noted metric is equivalent to 5D Minkowski space. This is a remarkable fact, and means that the matter in 4D comes in a sense from emptiness in 5D, in accordance with the STM approach.

The preceding example leads us to consider other cases where there is no ordinary matter but there is a classical Einstein vacuum whose energy density and pressure are measured by the cosmological constant Λ. The equation of state, with the constants absorbed, is $p = -\rho = -\Lambda/8\pi$ (see Chapter 1). To describe the vacuum in 5D STM requires a metric of the canonical type, a form of which we have met already in Eq. (2.6). Metrics of that type can describe many different physical systems, and in Chapter 7 we will consider the metrics for the inflationary universe and the one-body problem. The latter is of fundamental importance, especially for the classical tests of relativity in the solar system, so we quote it here:

$$dS^2 = \frac{\ell^2}{L^2}\left[\left(1 - \frac{2M}{r} - \frac{r^2}{L^2}\right)dt^2 - \frac{dr^2}{\left(1 - \frac{2M}{r} - \frac{r^2}{L^2}\right)} - r^2 d\Omega^2\right]$$
$$-d\ell^2 \quad . \tag{2.42}$$

Here M is the mass at the centre of the 3D geometry and $\Lambda = 3/L^2$. The sign of the cosmological constant changes if the sign of the extra term in the metric changes, so in STM theory it is possible in principle to determine whether the fifth dimension is spacelike or timelike.

2.5. Campbell's Theorem and the Field Equations

Embedding theorems can be classified as local and global in nature. We are primarily concerned with the former because our field equations are local. (The distinction is relevant, because global theorems are more difficult to establish, and involve boundary conditions which may be harder to satisfy.) There are several local embedding theorems pertinent to ND field theory, the most relevant of which is commonly attributed to Campbell [19]. He, however, only outlined a proof of this theorem in an academic treatise on differential geometry. It was subsequently studied and established by Magaard [20], resurrected by Tavakol and coworkers [21], and applied to gravitational theory by Seahra and Wesson [22]. The importance of Campbell's theorem is that it provides an algebraic method to proceed up or down the dimensionality ladder N of field theories like general relativity which are based on Riemannian geometry. Nowadays, it is possible to prove Campbell's theorem in short order using the lapse-and-shift technique associated with the ADM (Arnowitt, Deser and Misner) formulation of general relativity. This technique also provides insight to the connection between different versions of 5D gravity, such as induced-matter and membrane theory. We will have reason to appeal to Campbell's theorem at different places in our studies of 5D field theory. In the present section, we wish to draw on results by Ponce de Leon [16] and others, to give a brief account of the subject.

Campbell's theorem in succinct form says: Any analytic Riemannian space $V_n(s,t)$ can be locally embedded in a Ricci-flat Riemannian space $V_{n+1}(s+1,t)$ or $V_{n+1}(s,t+1)$.

We are here using the convention that the 'small' space has dimensionality n with coordinates running 0 to $n-1$, while the 'large' space has dimensionality $n+1$ with coordinates running 0 to n. The total dimensionality is $N = 1 + n$, and the main focus is on $N = 5$.

To establish the veracity of this theorem (in a heuristic fashion at least), and see its relevance (particularly to the theories considered before), consider an arbitrary manifold Σ_n in a Ricci-flat space V_{n+1}. The embedding can be visualized by drawing a line to represent Σ_n in a surface, the normal vector n^A to it satisfying $n \cdot n \equiv n^A n_A = \varepsilon = \pm 1$. If the vectors e_α^A form an appropriate basis and the extrinsic curvature of Σ_N is $K_{\alpha\beta}$, then the ADM constraints read

$$G_{AB} n^A n^B = -\frac{1}{2}\left(\varepsilon R_\alpha^\alpha + K_{\alpha\beta} K^{\alpha\beta} - K^2\right) = 0 \quad ,$$

$$G_{AB} e_\alpha^A n^B = \nabla_\beta K_\alpha^\beta - \partial_\alpha K = 0 \quad ,$$

(2.43)

where $K = g^{\alpha\beta} K_{\alpha\beta}$. These relations provide $n + 1$ equations for the $2 \times n(n+1)/2$ quantities $g_{\alpha\beta}$, $K_{\alpha\beta}$. Given an arbitrary geometry $g_{\alpha\beta}$

for Σ_n, the constraints therefore form an under-determined system for $K_{\alpha\beta}$, so infinitely many embeddings are possible. This implies that the embedding of a system of 4D equations like $G_{\alpha\beta} = 8\pi T_{\alpha\beta}$ in a system of 5D equations like $R_{AB} = 0$ is always possible.

This demonstration of Campbell's theorem can easily be extended to the case where V_{n+1} is a de Sitter space or anti-de Sitter space with an explicit cosmological constant, as in brane theory. Depending on the application, the remaining $n(n+1) - (n+1) = (n^2 - 1)$ degrees of freedom may be removed by imposing initial conditions on the geometry, physical conditions on the matter, or conditions on a boundary.

The last is relevant to Membrane theory with Z_2 symmetry. To see this, let us consider a fairly general line element with $dS^2 = g_{\alpha\beta}(x^\gamma, \ell)dx^\alpha dx^\beta + \varepsilon\, d\ell^2$ where $g_{\alpha\beta} = g_{\alpha\beta}(x^\gamma, +\ell)$ for $\ell \geq 0$ and $g_{\alpha\beta} = g_{\alpha\beta}(x^\gamma, -\ell)$ for $\ell \leq 0$ in the bulk [16]. Non-gravitational fields are confined to the brane at $\ell = 0$, which is a singular surface. Let the energy-momentum in the brane be represented by $\delta(\ell)S_{AB}$ (where $S_{AB}n^A = 0$) and that in the bulk by T_{AB}. Then the field equations read $G_{AB} = \kappa[\delta(\ell)S_{AB} + T_{AB}]$ where κ is a 5D coupling constant. The extrinsic curvature discussed above changes across the brane by an amount $\Delta_{\alpha\beta} \equiv K_{\alpha\beta}(\Sigma_{\ell>0}) - K_{\alpha\beta}(\Sigma_{\ell<0})$ which is given by the junction conditions. These imply

$$\Delta_{\alpha\beta} = -\kappa\left(S_{\alpha\beta} - \tfrac{1}{3}Sg_{\alpha\beta}\right) \quad . \tag{2.44}$$

But the $\ell = 0$ plane is symmetric, so

$$K_{\alpha\beta}(\Sigma_{\ell>0}) = -K_{\alpha\beta}(\Sigma_{\ell<0}) = -\frac{\kappa}{2}\left(S_{\alpha\beta} - \tfrac{1}{3}Sg_{\alpha\beta}\right) \quad . \tag{2.45}$$

This result can be used to evaluate the 4-tensor

$$P_{\alpha\beta} \equiv K_{\alpha\beta} - Kg_{\alpha\beta} = -\frac{\kappa}{2}S_{\alpha\beta} \quad . \tag{2.46}$$

However, $P_{\alpha\beta}$ is identical to the P-tensor $\left(\partial_4 g_{\alpha\beta} - g_{\alpha\beta}g^{\mu\nu}\partial_4 g_{\mu\nu}\right)/2\Phi$ of STM theory, which we noted above in Section 2.3. It obeys the field equations $\nabla_\beta P_\alpha^\beta = 0$, which are a subset of $R_{AB} = 0$. That is, the conserved tensor $P_{\alpha\beta}$ of STM theory is essentially the same as the total energy-momentum tensor in Z_2-symmetric M theory. Other correspondences can be established in a similar fashion.

Campbell's theorem is not a particularly strong statement, but it does have some notable implications. For example, the idea that the 4D equations $G_{\alpha\beta} = 8\pi T_{\alpha\beta}$ are contained in the 5D ones $R_{AB} = 0$ turns out not to be an assumption (as originally taken by the founders of STM theory) but a necessary consequence of Riemannian geometry. This alters somewhat our perception of the field equations derived in Section 2.3, for there is not really any question about whether they are 'right' or not: they are necessarily right given the starting metric in Eq. (2.7), as long as the underlying relations $R_{AB} = 0$ are right.

It may be convenient to restate the field equations of STM theory here in a compact form. Thus the analysis contained in Eqs. (2.7) – (2.34) can be summed up by the starting metric plus fifteen equations in groups of ten, four, and one, as follows:

$$dS^2 = g_{\alpha\beta}(x^\gamma, \ell)dx^\alpha dx^\beta + \varepsilon\Phi^2(x^\gamma, \ell)d\ell^2 \quad ; \qquad (2.47)$$

$$G_{\alpha\beta} = 8\pi T_{\alpha\beta} \quad ,$$

$$8\pi T_{\alpha\beta} \equiv \frac{\nabla_\beta \partial_\alpha \Phi}{\Phi} - \frac{\varepsilon}{2\Phi^2}\left\{ \frac{\partial_4\Phi\, \partial_4 g_{\alpha\beta}}{\Phi} - \partial_4\left(\partial_4 g_{\alpha\beta}\right) + \right.$$
$$g^{\lambda\mu}\,\partial_4 g_{\alpha\lambda}\,\partial_4 g_{\beta\mu} - \frac{1}{2}g^{\mu\nu}\,\partial_4 g_{\mu\nu}\,\partial_4 g_{\alpha\beta} +$$
$$\left. \frac{1}{4}g_{\alpha\beta}\left[\partial_4 g^{\mu\nu}\,\partial_4 g_{\mu\nu} + \left(g^{\mu\nu}\,\partial_4 g_{\mu\nu}\right)^2 \right] \right\} \quad ; \qquad (2.48)$$

$$\nabla_\beta P_\alpha^\beta = 0 \quad ,$$

$$P_\alpha^\beta \equiv \frac{1}{2\Phi}\left(g^{\beta\sigma}\,\partial_4 g_{\sigma\alpha} - \delta_\alpha^\beta g^{\mu\nu}\,\partial_4 g_{\mu\nu} \right) \quad ; \qquad (2.49)$$

$$\Box\Phi = -\frac{\varepsilon}{2\Phi}\left[\frac{\partial_4 g^{\lambda\beta}\,\partial_4 g_{\lambda\beta}}{2} + g^{\lambda\beta}\,\partial_4\left(\partial_4 g_{\lambda\beta}\right) - \frac{\partial_4\Phi g^{\lambda\beta}\,\partial_4 g_{\lambda\beta}}{\Phi} \right] \quad ,$$

$$\Box\Phi \equiv g^{\alpha\beta}\nabla_\beta \partial_\alpha \Phi \quad . \qquad (2.50)$$

The notation here is the same as in Section 2.3, where other useful relations may be found, including the 4D curvature scalar given by Eq. (2.25). The field equations (2.48) – (2.50) are general, insofar as they use only four of the five available degrees of coordinate freedom to remove the $g_{4\alpha}$ potentials. A few simple solutions of these equations were given in Section 2.4, and many others will emerge in subsequent chapters. To find exact solutions, it is necessary to restrict the general metric (2.47) in some way. Such restricted metrics are in the nature of gauges, and in Section 2.2 we met two such, namely the warp gauge of Eq. (2.5) and the canonical gauge of Eq. (2.6). Another example, which extends the former but has some consequences similar to the latter, is specified by Eq. (2.47) with $g_{\alpha\beta}(x^\gamma, \ell) = \exp(\ell\Phi/L)\bar{g}_{\alpha\beta}(x^\gamma)$ and $\Phi = \Phi(x^\gamma)$. As noted before, experience shows that the best way to approach the field equations is to look first at the pair of relations (2.50) and (2.49) for the scalar field Φ and the P-tensor, and then consider the Einstein-like equations (2.48). However, the reader may well have other ideas, and is invited to try them.

2.6. Conclusion

The 5D field equations $R_{AB} = 0$ are deceptively simple. They resemble, of course, the 4D ones $R_{\alpha\beta} = 0$ which are verified by the classical tests of general relativity. But the 5D ones, as we have seen in the present chapter, also contain matter. In fact, they contain the traditional forms of matter, including vacuum, plus some new forms which are yet to be observed. Space-Time-Matter theory not only unifies the gravitational, electro-magnetic and scalar fields, but also unifies these with their sources. It fulfills Einstein's desire to give matter a geometrical origin.

The choice between describing nature in 5D or 4D terms is thus equivalent to deciding whether we wish to treat fields and their matter as separate, or consider them as parts of the same thing. The distinction could be approached as an issue of philosophy. Bertrand Russell, for example, opined that the stuff of physics is what obeys its equations, a view which no doubt appealed to him because equations are based on logic and he believed logic to be the basis of everything that the human mind can understand. Sir Arthur Eddington meant something similar when he summed up his view of science by saying "To put the conclusion crudely

– the stuff of the world is mind-stuff". However, the question of fields versus sources can also be regarded as a practical one. Take, as something familiar to every working physicist, Poisson's equation: $\nabla^2 \phi = 4\pi G\rho$. We are used to seeing it in this form, which suggests that the matter with density ρ is somehow responsible for the gravitational field as requested by its potential ϕ. But Poisson's equation is just as valid when written as $4\pi G\rho = \nabla^2 \phi$, with the reverse interpretation. The situation is actually very simple: the field and the matter are parts of the same thing.

STM theory formalizes this idea. If we take one of the components of the equation $R_{AB} = 0$ and write out all of its terms, it is always possible to separate them into these which do not involve the extra dimension and those which do. Putting these on the left and right sides of an equals sign produces a relation which conventionally we call the field and its matter, as in $\nabla^2 \phi = 4\pi G\rho$. However, this division is arbitrary. Einstein recognized this, and in what follows we follow his insight, with the fifth dimension as a unifying theme.

References

[1] Kaluza, T., *Sitz. Preuss. Akad. Wiss.* **33**, 966 (1921).

[2] Klein, O., *Z. Phys.* **37**, 895 (1926).

[3] Overduin, J.M., Wesson, P.S., *Phys. Reports* **283**, 303 (1997).

[4] Wesson, P.S., *Space-Time Matter*, 2nd. ed. (World Scientific, Singapore, 2007).

[5] Wesson, P.S., *Five-Dimensional Physics* (World Scientific, Singapore, 2006).

[6] Overduin, J.M., Wesson, P.S., *The Light/Dark Universe* (World Scientific, Singapore, 2008).

[7] Will, C.M., *Theory and Experiment in Gravitational Physics* (Cambridge U. Press, Cambridge, 1993).

[8] Wesson, P.S., Ponce de Leon, J., *J. Math. Phys.* **33**, 3883 (1992).

[9] Wesson, P.S., *Astrophys. J.* **394**, 19 (1992).

[10] Wesson, P.S., *Phys. Lett.* **B276**, 299 (1992).

[11] Mashhoon, B., Liu, H., Wesson, P.S., *Phys. Lett.* **B331**, 305 (1994).

[12] Randall, L., Sundrum, R., *Mod. Phys. Lett.* **A13**, 2807 (1998).

[13] Randall, L., Sundrum, R., *Phys. Rev. Lett.* **83**, 4690 (1999).

[14] Arkani-Hamed, N., Dimopoulos, S., Dvali, G., *Phys. Lett.* **B429**, 263 (1998).

[15] Arkani-Hamed, N., Dimopoulos, S., Dvali, G., *Phys. Rev.* **D59**, 086004 (1999).

[16] Ponce de Leon, J., *Mod. Phys. Lett.* **A16**, 2291 (2001).

[17] Einstein, A., Bergmann, P., *Ann. Math.* **39**, 683 (1938).

[18] Ponce de Leon, J., *Gen. Rel. Grav.* **20**, 539 (1988).

[19] Campbell, J.E., *A Course of Differential Geometry* (Clarendon, Oxford, 1926).

[20] Halpern, P., *The Great Beyond* (Wiley, Hoboken, 2004); Magaard, L., Ph.D. Thesis, Kiel (1963).

[21] Rippl, S., Romero, C., Tavakol, R., *Class. Quant. Grav.* **12**, 2411 (1995). Romero, C., Tavakol, R., Zalaletdinov, R., *Gen. Rel. Grav.* **28**, 365 (1996). Lidsey, J.E., Romero, C., Tavakol, R., Rippl, S., *Class. Quant. Grav.* **14**, 865 (1997).

[22] Seahra, S.S., Wesson, P.S., *Class. Quant. Grav.* **20**, 1321 (2003).

Chapter 3

Cosmology in 5D

3.1. Introduction

Five-dimensional Space-Time-Matter theory is guaranteed by construction to contain everything found in four-dimensional general relativity. We saw in Chapter 1 that there are models in Einstein's theory which describe the inflationary and radiation-dominated eras of the early universe; and while the precise nature of dark energy is uncertain, models with a cosmological constant also describe with some success the current behaviour, where the expansion appears from supernova data to be accelerating. However, under a broad range of conditions, the majority of cosmological models based on Einstein's theory begin with a big bang. This singular event is viewed positively by some people as a birth event for the universe, and negatively by others because Einstein's equations lose their usefulness at that point. Since adding a dimension broadens the scope of the physical theory, it is possible in STM to have different accounts for the very early universe, including ones without a big bang. In order to present the material of 5D cosmology, we therefore give in this chapter a series of 5D alternatives to the conventional big bang.

The phrase "big bang" was coined by Fred Hoyle, who applied it in a contemptuous manner to an idea he thought intrinsically illogical. Hoyle with Bondi and Gold preferred as a basis for cosmology the idea that the observable universe is in a steady state, where matter is created by some agency (perhaps by a scalar field), which counters the dilution caused by the explosion of the galaxies, keeping the average density the same not only at all places in space but also at all moments in time (see Chapter 1). This idea proved typical of others in the sociologically progressive England of the 1960s: elegant but ultimately misguided. Steady-state cosmology was abandoned after it proved to be in conflict with galaxy counts and lacked a plausible explanation for the origin and isotropy of the microwave background.

Realizing that some kind of special physical event was needed at time zero, Hoyle in 1975 suggested another alternative to the big bang, namely a transition from negative to positive particle mass in the context of a 4D conformally-invariant theory [1]. Other workers subsequently came up with different alternatives. For example, Vilenkin suggested that it could have been a quantum tunneling event [2]; while Wesson following Bonnor showed that it was possible to create matter from Minkowski space in a blip-like event [3].

Developments in 5D physics have made it possible to interpret these and other 4D attempts to circumvent the big bang in a new way. Thus, Ponce de Leon found a class of 5D cosmologies which reduce to the Friedmann-Lemaître-Robertson-Walker ones on hypersurfaces where the extra coordinate is constant [4]. Wesson and Seahra showed that while these cosmologies are curved in 4D, they are actually flat in 5D, implying that the initial singularity is merely a coordinate effect. Further work showed this to be true for other cosmological and astrophysical solutions as well [6]. In what follows, we will have a closer look at the Ponce de Leon cosmologies, and then consider other solutions of the 5D field equations which provide alternatives to the big bang.

3.2. Flat 5D Universes

The standard cosmological models in 5D – the analogs of the FLRW ones in 4D – were derived by Ponce de Leon [4]. They have been much studied, and their consequences are mentioned at numerous places in this volume. They can be given a pictorial representation by a combination of algebra and computer work [5], and we follow this approach here because it gives insight to the nature of the 4D singularity.

The models are commonly written in coordinates $x^0 = t$, $x^{123} = r\theta\phi$ and $x^4 = \ell$. (As usual, we absorb the speed of light and the gravitational constant through a choice of units which renders them unity.) The line element is given by

$$dS^2 = \ell^2 dt^2 - t^{2/\alpha}\ell^{2/(1-\alpha)}(dr^2 + r^2 d\Omega^2)$$
$$- \frac{\alpha^2 t^2}{(1-\alpha)^2} d\ell^2 \ , \tag{3.1}$$

where $d\Omega^2 \equiv (d\theta^2 + \sin^2\theta\, d\phi^2)$. The dimensionless parameter α is related to the properties of matter.

The latter can be obtained using the technique outlined in Chapter 2, where we used Campbell's theorem to embed 4D general relativity with Einstein's equations $G_{\alpha\beta} = 8\pi T_{\alpha\beta}$ ($\alpha,\beta = 0,123$) in an apparently empty 5D manifold with field equations $R_{AB} = 0$ ($A, B = 0,123,4$). Here the effective or induced-matter energy-momentum tensor can be taken as that of a perfect fluid with density ρ and pressure p. Then the class of solutions (3.1) fixes these quantities via

$$8\pi\rho = \frac{3}{\alpha^2\tau^2} \ , \quad 8\pi p = \frac{2\alpha-3}{\alpha^2\tau^2} \ , \tag{3.2}$$

where $\tau = \ell t$ is the proper time. The equation of state is $p = (2\alpha/3 - 1)\rho$. For $\alpha = 3/2$, the scale factor of Eq. (3.1) varies as $t^{2/3}$, the density and pressure of Eq. (3.2) are $\rho = 1/6\pi\tau^2$ with $p = 0$, and we have the standard $k = 0$ dust model for the late universe. For $\alpha = 2$, the scale factor varies as $t^{1/2}$, $\rho = 3/32\pi\tau^2 = 3p$ and we have the standard $k = 0$ radiation model for the early universe. Cases with $\alpha < 1$ describe models that expand more quickly than the standard FLRW ones and have inflationary equations of state (see below). It should be noted that ρ and p in Eq. (3.2) refer to the total density and pressure respectively. These could be split into multiple components, including visible matter, dark matter, and possible vacuum (scalar) fields. The last may include a contribution from a time-varying cosmological 'constant' [7] of the type suggested by some data on the dynamics of galaxies and the age of the universe.

Physically, the Ponce de Leon cosmologies are reasonable. Mathematically, they are flat in three dimensions, curved in four dimensions, and flat in five dimensions. This means that Eq. (3.1) in coordinates $(t, r, \theta, \phi, \ell)$ is equivalent to 5D Minkowski space in some other coordinates (T, R, θ, ϕ, L) with line element

$$dS^2 = dT^2 - (dR^2 + R^2 d\Omega^2) - dL^2 \ . \tag{3.3}$$

This does not have a big bang, but the four-dimensional part of Eq. (3.1) *does* (the 4-geometry is singular for $t \to 0$). A situation similar to this occurs in general relativity with the Milne model. In many books this is presented as one of the FLRW class with negative spatial curvature, but a fairly simple coordinate transformation makes the metric a 4D Minkowski space, and accordingly it is devoid of matter [8]. The coordinate transformation between Eq. (3.1) and Eq. (3.3) for the corresponding 5D case is not simple. It is given by

$$T(t, r, \ell) = \frac{\alpha}{2} \left[\left(1 + \frac{r^2}{\alpha^2}\right) t^{\left(\frac{1}{\alpha}\right)} \ell^{\left(\frac{1}{1-\alpha}\right)} - \frac{t^{\left(\frac{2\alpha-1}{\alpha}\right)} \ell^{\left(\frac{1-2\alpha}{1-\alpha}\right)}}{1-2\alpha} \right] ,$$

$$R(t, r, \ell) = r t^{\left(\frac{1}{\alpha}\right)} \ell^{\left(\frac{1}{1-\alpha}\right)} , \tag{3.4}$$

$$L(t, r, \ell) = \frac{\alpha}{2} \left[\left(1 - \frac{r^2}{\alpha^2}\right) t^{\left(\frac{1}{\alpha}\right)} \ell^{\left(\frac{1}{1-\alpha}\right)} + \frac{t^{\left(\frac{2\alpha-1}{\alpha}\right)} \ell^{\left(\frac{1-2\alpha}{1-\alpha}\right)}}{1-2\alpha} \right] .$$

We have made an extensive study of these relations, in order to better understand the nature of the big bang.

The 4D physics occurs in the FLRW-like coordinates (t, r) of Eq. (3.1) on a hypersurface $\ell = \ell_0$ (the angular variables play no physical role and may be suppressed). The 4D models may, however, be regarded as embedded in a flat space with the coordinates (T, R, L) of Eq. (3.3) and viewed therefrom. With the help of Eq. (3.4), we can thus obtain pictures of the 4D models and study the structure of their singularities. This can be done for a range of the assignable parameters (α, ℓ_0). As an aid to visualization, we let r and R run over positive and negative values so that the images are symmetric about $R = 0$ (if it is desired to have $r, R > 0$, then one of the symmetric halves may be deleted). To the same end, we add lines of constant t that intersect the $R = 0$ plane orthogonally and lines of constant r that run parallel to the symmetry plane at $R = 0$. The models grow in ordinary space as they evolve in time. We present informative cases which are illustrated in the accompanying figures, as follows:

- *Model I* ($\alpha = 3/2, \ell_0 = 1$). This is the standard $k = 0$ model for the late universe. By Eq. (3.1) it has a scale factor that varies as $t^{2/3}$, and by Eq. (3.2) it has $p = 0$. The shape is parabolic, and lines of constant r meet at a pointlike big bang at $T = R = L = 0$.
- *Model II* ($\alpha = 1/30, \ell_0 = 1$). This is an inflationary $k = 0$ model for the very early universe. By Eq. (3.1) it has a scale factor that varies as t^{30}, and by Eq. (3.2) it has $(\rho + 3p) < 0$, so what is sometimes called the gravitational density is negative and powers a strong acceleration. Comoving trajectories converge to a point arrived at by following the null ray $T + L = R = 0$ into the past, towards past null infinity.
- *Model III* ($\alpha = 1/3, \ell_0 = 20, 40, 60$). This is a set of inflationary $k = 0$ models with scale factors that vary as t^3, $(\rho + 3p) < 0$ and a common big bang at past null infinity.

Figures 3.1 and 3.2 reproduce known physics for the standard spatially-flat FLRW models while adding a new perspective. Also, our figures for classical inflationary models are strikingly similar to those generated by computer for a stochastic theory of inflation based on quantum field theory [9]. In theories of the latter type, the rest masses of particles are basically zero and become finite through a mechanism involving the Higgs field. In 5D classical theory, it has been argued that the Higgs potential is related to the g_{44}-component of the metric tensor [6]. Alternatively, this component may be related to the effective 4D cosmological constant [7]. In either case, we see from Eq. (3.1) that this factor is time-dependent, which raises the possibility of testing such models using particle masses and gravitational lensing.

In general, the dynamics of models described by Eq. (3.1) may be studied by solving the 5D geodesic equation. We will give this detailed consideration elsewhere, but note here some results which follow if we use the 5D proper time S in Eq. (3.1) to characterize the motion. Thus the 5D geodesic gives the 5-velocities $U^A = dx^A/dS$ as $U^i = 0$ ($i = 1,2,3$), with $U^0 = \mp\alpha(2\alpha - 1)^{-1/2}\ell^{-1}$ and $U^4 = \pm(1 - \alpha)^2\alpha^{-1}(2\alpha - 1)^{-1/2}t^{-1}$. There is no motion in 3-space, and the galaxies are static with respect to

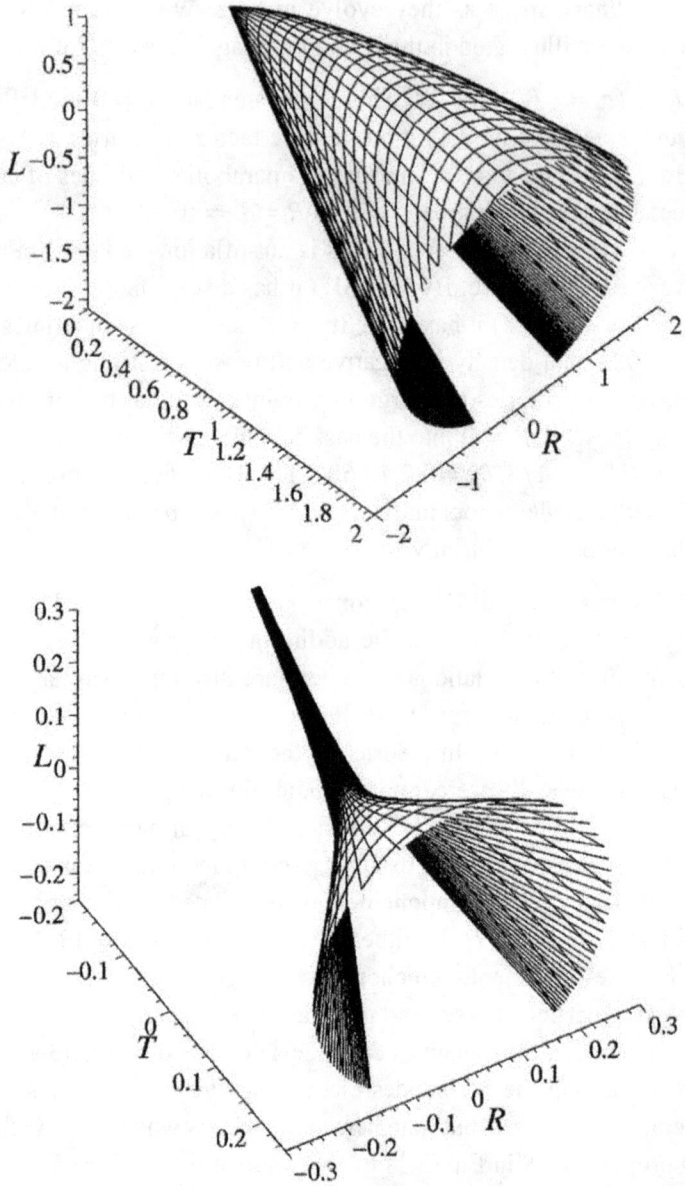

Figure 3.1: Hypersurfaces in 5D corresponding to spatially-flat FLRW cosmologies in 4D. The upper case is the standard dust model, the lower case is an inflationary model, as discussed in the text.

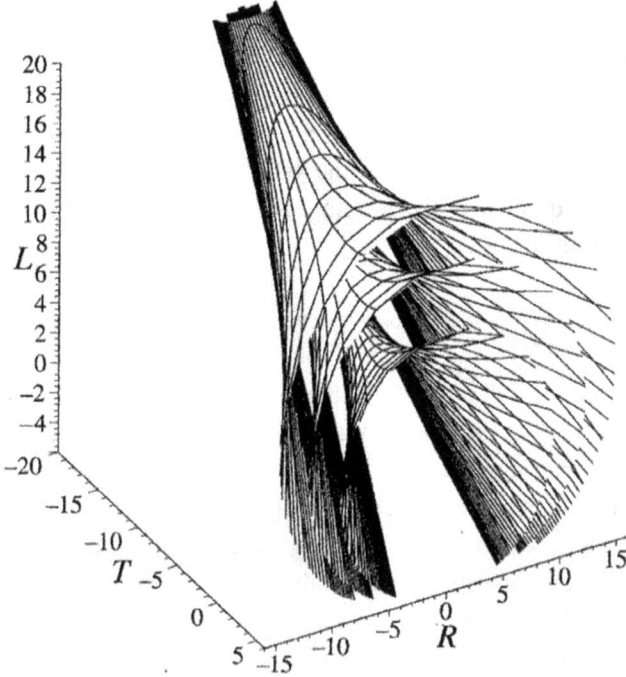

Figure 3.2: Hypersurfaces in 5D corresponding to spatially-flat, inflationary FLRW cosmologies in 4D. The outer surface has $\ell_0 = 60$, the middle surface has $\ell_0 = 40$ and the inner surface has $\ell_0 = 20$, as discussed in the text.

each other because the coordinates (t, r, ℓ) in Eq. (3.1) are designed to be spatially comoving. This is the same prescription as used in most presentations of the 4D Robertson-Walker metric [8]. The motions detected spectroscopically in observational cosmology refer to a noncomoving frame. In 5D, the coordinate transformation to $\bar{t} = \ell t$, $\bar{r} = t^{1/\alpha} r$ and $\bar{\ell} = At^A \ell$ (where A is a constant introduced for the purely algebraic purpose of distinguishing $\bar{t}$ from $\bar{\ell}$) results in $\bar{U}^0 = \mp(2\alpha - 1)^{1/2}\alpha^{-1}, \bar{U}^4 = 0, \bar{U}^\theta = \bar{U}^\phi = 0$ and $\bar{U}^1 = \mp(2\alpha - 1)^{-1/2}\bar{r}/\bar{t}$. The last expression is just Hubble's law.

We see that the 5D Ponce de Leon models defined by Eq. (3.1) have the same law for galaxy motions as the standard 4D FLRW models, as well as the same expressions for the density and pressure, Eqs. (3.2).

However, the line element (3.1) is related to the 5D Minkowski one (3.3) by the coordinate transformations (3.4). This remarkable fact may be confirmed by symbolic computing software. There may be a singularity in the matter-filled and curved 4D space, but one does not exist in the empty and flat 5D space. In other words, the 4D big bang is due to an unfortunate choice of coordinates in a smooth 5D manifold. To this extent, it is something of an illusion.

3.3. The Singularity as a Shock Wave

In this section, we will summarize the properties of another exact solution of the 5D field equations whose Riemann-Christoffel tensor obeys $R_{ABCD} = 0$, meaning that it is flat [10, 11]. However, the new solution depends only on the combined variable $u \equiv (t - \ell)$, so it describes a wave. Solutions of this type in Newtonian hydrodynamics where the density ρ and/or pressure p change abruptly are called shock waves. We can evaluate these properties of matter as before, by studying the 4D Einstein equations $G_{\alpha\beta} = 8\pi T_{\alpha\beta}$ ($\alpha, \beta = 0,123$), which are contained in the 5D field equations $R_{AB} = 0$ ($A, B = 0,123,4$). The properties of the 5D solution imply that we can view the 4D singularity as a kind of shock wave.

The solution has a 5D line element given by

$$dS^2 = b^2 dt^2 - a^2(dr^2 + r^2 d\Omega^2) - b^2 d\ell^2 \quad,$$

$$a = (hu)^{\frac{1}{2+3\alpha}} \quad, \tag{3.5}$$

$$b = (hu)^{-\frac{1+3\alpha}{2(2+3\alpha)}} \quad.$$

The notation here is the same as in the preceding section, and the solution may be confirmed using the algebra of Chapter 1 or by computer. It depends on two constants, h and α. The first has the physical dimensions of inverse length or time, and is related to Hubble's parameter (see below). The second is dimensionless, and is related to the properties of matter. There is an associated equation of state, and after some algebra we find

$$p = \alpha\rho \quad , \quad 8\pi\rho = \frac{3h^2}{(2+3\alpha)^2} a^{-3(1+\alpha)} \quad . \tag{3.6}$$

We see that $\alpha = 0$ corresponds to the late (dust) universe while $\alpha = 1/3$ corresponds to the early (radiation) universe.

To elucidate the physical properties of the solution, it is instructive to change from the coordinate time t to the proper time T. This is defined by $dT = b \, dt$, so

$$T = \frac{2}{3}\left(\frac{2+3\alpha}{1+\alpha}\right)\frac{1}{h}(hu)^{\frac{3}{2}\left(\frac{1+\alpha}{2+3\alpha}\right)} \quad . \tag{3.7}$$

The 4D scale factor which determines the dynamics of the model by Eqs. (3.5) and (3.7) is then

$$a(T) = \left[\frac{3}{2}\left(\frac{1+\alpha}{2+3\alpha}\right)hT\right]^{\frac{2}{3(1+\alpha)}} \quad . \tag{3.8}$$

For $\alpha = 0$, $a(T) \propto T^{2/3}$ as in the standard (Einstein-de Sitter) dust model. For $\alpha = 1/3$, $a(T) \propto T^{1/2}$ as in the standard radiation model. The value of Hubble's parameter is given by

$$H \equiv \frac{1}{a}\frac{\partial a}{\partial T} = \frac{1}{a}\frac{\partial a}{\partial t}\frac{dt}{dT} = \frac{h}{(2+3\alpha)}(hu)^{-\frac{3}{2}\left(\frac{1+\alpha}{2+3\alpha}\right)}$$

$$= \frac{2}{3(1+\alpha)T} \quad . \tag{3.9}$$

For $\alpha = 0$ and $1/3$, Eq. (3.9) shows that H has its standard values in terms of the proper time. We can also convert the density in Eq. (3.6) from t to T using Eq. (3.7), giving

$$8\pi\rho = \frac{4}{3(1+\alpha)^2 T^2} \quad . \tag{3.10}$$

For $\alpha = 0$ we have $\rho = 1/6\pi T^2$, and for $\alpha = 1/3$ we have $\rho = 3/32\pi T^2$, the standard FLRW values. Thus, the 5D solution (3.5) contains 4D dynamics and 4D matter that are the same as in the standard 4D cosmologies for the late and early universe.

However, while the 5D approach does no violence to the 4D one, it adds significant insight. The big bang occurs in proper time at $T = 0$ by Eq. (3.10); but it occurs in coordinate time at $a = 0$ or $u = t - \ell = 0$ by Eqs. (3.6) and (3.5). The field equations $R_{AB} = 0$ are fully covariant, so any choice of coordinates is valid. Therefore, we can interpret the physically-defined big bang either as a singularity in 4D or as a hypersurface $t = \ell$ that represents a plane wave propagating in 5D.

Some comments are in order about the shock-wave solution (3.5), the flat-universe solution (3.1) and certain other cosmological solutions in the literature [6]. These all have $R_{ABCD} = 0$, and it is possible to make a systematic study of these equations [12]. However, it is *not* usually possible to find coordinate transformations between solutions, or show the explicit transformation to 5D Minkowski space like that given in Eq. (3.4), because of the level of complexity involved. Further, a given 5D solution may have different 4D interpretations. This is because the group of 5D coordinate transformations $x^A \rightarrow x^A(x^B)$ is broader that the 4D group $x^\alpha \rightarrow x^\alpha(x^\beta)$, so x^4-dependent transformations are mathematically equivalent in 5D but physically *non*-equivalent in 4D. Consequences of this are discussed elsewhere as appropriate, but it also has implications for the physics of the big bang. Consider, for example, the cosmic microwave background. In the conventional 4D picture, this is thermalized in the hot, dense fireball that followed the singularity and cooled by the expansion of the universe. But in another coordinate frame, such as a 5D Minkowski one, there is no big bang and no expansion. We infer that in this frame the background radiation must be thermalized by some other mechanism. Hoyle has suggested one such, in which particle masses vary with epoch, in such a way that their Thomson cross-section diverged for small times and caused efficient scattering [1]. Coordinate-dependent physics may sound strange in a gravitational context, but it is actually common in quantum theory [13], where non-covariant gauges are employed to solve practical problems. Gauge-dependent 4D physics is essentially unavoidable in 5D relativity.

3.4. Bounce versus Bang

It has sometimes been speculated that 4D FLRW models with positive 3D curvature might, after their expansion phase, collapse to a 'big crunch', from which they might re-emerge. However, this idea owes more to a belief in reincarnation than to physics, where it cannot be proved because it is impossible to integrate Einstein's equations through a second or (nth) singularity. It has also occasionally been suggested that an FLRW model with a genuine bounce, where there is a contraction to a minimum but finite scale followed by an expansion, might serve to describe the real universe. But this idea encounters problems when confronted with observational data, notably on the ages of globular clusters and the redshifts of quasars [7, 14-16]. Furthermore, supernovae data imply that the universe on the largest scales is accelerating, implying a significant positive cosmological constant or some other dark form of energy with similar consequences. Indeed, observations now imply that 70% of the universe consists of vacuum energy. In STM theory, the natural way to handle the cosmological constant or a similar vacuum tensor is by the 5D canonical metric, which we will examine in detail in Chapter 7. In the present section, we wish to look at a type of cosmology which uses the canonical metric and features a big bounce.

Cosmology based on the 5D canonical metric was developed by Liu and his coworkers in the years 1994-2006. The actual metric was introduced by Mashhoon, Liu and Wesson in 1994 [17]. It consists of a quadratic factor in the extra coordinate multiplied onto 4D spacetime, plus an extra flat piece. In the limit where the 4D part does not depend on the extra coordinate (beyond the quadratic factor), the canonical metric gives back all 4D spacetimes with a cosmological constant. In this regard, it is similar to the warp metric of Membrane theory. Cosmological field equations for the canonical metric were isolated by Liu and Mashhoon, who pointed out that the most logical interpretation of the extra coordinate was in terms of particle mass, which must be variable over cosmic time [18]. They also commented on how observations should be interpreted in such a variable-mass world, which must be dominated on large scales by the cosmological constant or vacuum energy [19]. Then in 2001, Liu and Wesson found that the 5D field equations for the canonical metric admitted an event which, unlike the big bang, could be termed a big bounce [20]. In fact, the models have a bang *and* a bounce, with the latter being the later event, and evolve into vacuum-dominated states at late times in accordance with the recently-discovered properties of dark energy. The field equations for canonical cosmological models can be reduced to a pair of relations which are the analogs of the Friedmann-Lemaître equations of general relativity (see Chapter 1). These were subsequently investigated by Liu and coworkers, particularly in regard to the density and pressure of the matter verses the vacuum [21, 22]. Eventually, a model was found with matter, radiation and vacuum components, such that the former two decreased with time while the latter increased with time [23]. In fact, using a phenomenological expression of Wetterich for the equation of state of the dark energy, the model was fully determined in its parameters and show to have properties which when plotted as a function of redshift were in agreement with the observational data available in 2006. Also in that year, a study was made of possible radiation from isolated objects described by the canonical/warp metric, combining the field equations with the Klein-Gordon equation to see if there were other observable consequences of the 5D approach [24]. The results of this last study were somewhat intricate, and for practical application would require new observations of collapsed astrophysical objects such as black holes. By contrast, the possible existence of a big bounce is a striking result, which we proceed to investigate.

To obtain ordinary matter as well as vacuum energy from the canonical metric, it is necessary to add some dependence on $x^4 = \ell$ beyond the typical quadratic term [20]. It turns out that the most convenient form of the line element, and one that satisfies $R_{AB} = 0$, is

$$dS^2 = B^2 dt^2 - A^2 \left(\frac{dr^2}{1-kr^2} + r^2 d\Omega^2 \right) - d\ell^2 \quad ,$$

$$A^2 = (\mu^2 + k)\ell^2 + 2v\ell + \frac{v^2 + K}{\mu^2 + k} \quad , \tag{3.11}$$

$$B = \frac{1}{\mu} \frac{\partial A}{\partial t} \equiv \frac{\dot{A}}{\mu} \quad .$$

Here $\mu = \mu(t)$ and $v = v(t)$ are arbitrary functions, k is the 3D curvature index ($k = \pm 1, 0$) and K is a constant. After a lengthy calculation, we find that the 5D Kretschmann invariant takes the form

$$I = R_{ABCD}R^{ABCD} = \frac{72K^2}{A^8} \ , \tag{3.12}$$

which shows that K determines the curvature of the 5D manifold. From Eqs. (3.11) we see that the form of $B\,dt$ is invariant under an arbitrary transformation $t = t(\bar{t})$. This gives us the freedom to fix one of the two arbitrary functions $\mu(t)$ and $v(t)$ without changing the basic solutions. The other arbitrary function can be seen to relate to the 4D properties of matter, which we now discuss.

The 4D line element is

$$ds^2 = g_{\alpha\beta}dx^{\alpha}dx^{\beta} = B^2 dt^2 - A^2\left(\frac{dr^2}{1-kr^2} + r^2 d\Omega^2\right) . \tag{3.13}$$

This has the Robertson-Walker form which underlies the standard FLRW models, and allows us to calculate the non-vanishing components of the 4D Ricci tensor:

$$^4R_0^0 = -\frac{3}{B^2}\left(\frac{\ddot{A}}{A} - \frac{\dot{A}\dot{B}}{AB}\right) \ ,$$

$$^4R_1^1 = {}^4R_2^2 = {}^4R_3^3 = -\frac{1}{B^2}\left[\frac{\ddot{A}}{A} + \frac{\dot{A}}{A}\left(\frac{2\dot{A}}{A} - \frac{\dot{B}}{B}\right) + 2k\frac{B^2}{A^2}\right] . \tag{3.14}$$

Now from Eq. (3.11) we have

$$B = \frac{\dot{A}}{\mu} \ , \quad \dot{B} = \frac{\ddot{A}}{\mu} - \frac{\dot{A}\dot{\mu}}{\mu\mu} \ .$$

Using these expressions in Eq. (3.14), we can eliminate B and $\dot{B}$ to give

$$^4R_0^0 = -\frac{3\mu\dot{\mu}}{A\dot{A}} \ ,$$

$$^4R_1^1 = {}^4R_2^2 = {}^4R_3^3 = -\left[\frac{\mu\dot{\mu}}{A\dot{A}} + \frac{2(\mu^2+k)}{A^2}\right] . \tag{3.15}$$

The 4D Ricci scalar is then found to be

$$^4R = -6\left(\frac{\mu\dot{\mu}}{A\dot{A}} + \frac{\mu^2+k}{A^2}\right) \ .$$

(3.16)

Eqs. (3.15) and (3.16) together allow us to construct the 4D Einstein tensor $G_\beta^\alpha \equiv {}^4R_\beta^\alpha - \delta_\beta^\alpha\, {}^4R/2$, with nonvanishing components

$$G_0^0 = \frac{3(\mu^2+k)}{A^2} \ ,$$

$$G_1^1 = G_2^2 = G_3^3 = \frac{2\mu\dot{\mu}}{A\dot{A}} + \frac{\mu^2+k}{A^2} \ .$$

(3.17)

These expressions in turn give the components of the induced energy-momentum tensor, since Einstein's equations $G_\beta^\alpha = 8\pi T_\beta^\alpha$ hold.

Let us suppose that the induced matter is a perfect fluid with density ρ and pressure p, moving with a 4-velocity $u^\alpha \equiv dx^\alpha/ds$, plus a cosmological term whose nature is to be determined. Then we have

$$G_{\alpha\beta} = 8\pi\left[(\rho+p)u_\alpha u_\beta + (\Lambda/8\pi - p)g_{\alpha\beta}\right] \ .$$

(3.18)

As in the FLRW models, we can take the matter to be comoving in three dimensions, so $u^\alpha = (u^0, 0, 0, 0)$ and $u^0 u_0 = 1$. Then Eqs. (3.18) and (3.17) yield

$$8\pi\rho + \Lambda = \frac{3(\mu^2+k)}{A^2} \ ,$$

$$8\pi p - \Lambda = -\frac{2\mu\dot{\mu}}{A\dot{A}} - \frac{\mu^2+k}{A^2} \ .$$

(3.19)

These are the analogs for our solution (3.11) of the Friedmann-Lemaître equations for the FLRW solutions. As before, we are free to choose an equation of state, which we take to be the isothermal one,

$$p = \gamma\rho \ .$$

(3.20)

Here γ is a constant, which for ordinary matter takes values between 0 (for dust) 0 and 1/3 (for radiation or ultrarelativistic particles). Substituting Eq. (3.20) into Eqs. (3.19), we can isolate the density of matter and the cosmological term:

$$8\pi\rho = \frac{2}{1+\gamma}\left(\frac{\mu^2+k}{A^2} - \frac{\mu\dot{\mu}}{A\dot{A}}\right) \; ,$$

$$\Lambda = \frac{2}{1+\gamma}\left[\left(\frac{1+3\gamma}{2}\right)\left(\frac{\mu^2+k}{A^2}\right) + \frac{\mu\dot{\mu}}{A\dot{A}}\right] \; . \tag{3.21}$$

In these relations, $\mu = \mu(t)$ is still arbitrary and $A = A(t,\ell)$ is given by Eqs. (3.11). The matter density therefore has a wide range of forms, and the cosmological constant is in general variable.

Let us now consider singularities of the manifold described by Eqs. (3.11). Since this is 5D Ricci-flat, we have $R = 0$ and $R^{AB}R_{AB} = 0$. The third 5D invariant is given by Eq. (3.12), from which we see that the condition $A = 0$ (with $K \neq 0$) corresponds to a 5D singularity. This is a physical singularity, and as in general relativity, can naturally be termed a 'big bang.' However, from Eqs. (3.16) and (3.21), we also see that if $\dot{A} = 0$, then all of the 4D quantities ^4R, ρ and Λ diverge. But while this defines a kind of 4D singularity, the 5D curvature invariant (3.12) does *not* diverge. This is a second kind of 4D singularity, associated with the minimum in the 3D scale factor A, and can reasonably be described as a big bounce. We further note from Eq. (3.11) that if $\dot{A} = 0$ then $B = 0$ (assuming $\mu \neq 0$), and the time part of the 4D line element vanishes. To sum up: the manifold (3.11) has a 5D geometrical singularity associated with $A = 0$ and a 4D matter singularity associated with $B = 0$, which we term respectively a big bang and a big bounce.

The physics associated with the bounce, and plots of the 3D scale factor $A = A(t,\ell)$ as a function of the time t for various values of the extra coordinate ℓ, were studied by Liu and Wesson [20]. They put $k = 0$ in Eq. (3.11) on the basis of observational evidence, and chose functions $\mu(t)$ and $\nu(t)$ for algebraic convenience. They found that, typically, the form of the scale factor $A(t,\ell)$ is not symmetric around the minimum or time of the bounce. This implies that in 5D models of this type, there is a 4D production of entropy and/or matter around the bounce. Other work on

Eq. (3.11) and its associated relations has been done, particularly in relation to membrane tension in M theory and phase transitions in STM theory ([25-29]; for a review see Ref. [30]). That the metric (3.11) has more than one application is due to the generality of its algebra. And we will see in the following section that there is one more application that is truly remarkable.

3.5. The Universe as a 5D Black Hole

The concept of a 4D black hole is now so familiar that it is automatically associated with a central singularity, surrounded by an event horizon which depends on the mass M at the centre of a spherically-symmetric 3D space which is asymptotically flat. The latter property and others imply that the Schwarzschild solution is – up to coordinate transformations – unique. This is embodied in Birkhoff's theorem, which plays a significant role in Einstein's general theory of relativity. By comparison, the concept of a 5D "black hole" is considerably more complicated, due to the extra degrees of freedom introduced by the fifth coordinate. The 5D analogs are known as solitons, and will be examined in Chapter 5. There we will see that there are actually several different kinds of solitons. Thus Birkhoff's theorem, in its conventional form, fails in 5D. Indeed, it is unwise to carry over preconceptions about black holes from 4D to 5D. We will see below that it is more advisable to consider a topological 5D black hole, defined by the symmetries of its metric, and work its properties without presumptions. Specifically, our aim in this section is to consider a general 5D metric of the black-hole type, and show that it is isometric to that of the cosmologies treated in the previous section ([31, 32] see also Ref. [33]). That is, we wish to ponder the possibility that the universe may be a 5D black hole.

As usual, we let upper-case Latin indices run over 0–4 and lower-case Greek indices run over 0–3. We use time and spherical-polar spatial coordinates plus a length ℓ, so $x^A = (t, r, \theta, \phi, \ell)$. Then $d\Omega_2^2 \equiv d\theta^2 + \sin^2\theta d\phi^2$ is the measure on a 2D spherical shell. The topological black-hole class of 5D solutions is given by

$$dS^2 = hdT^2 - h^{-1}dR^2 - R^2 d\Omega_3^2 \ . \tag{3.22}$$

Here T is the time and $h = h(R)$ is a function of the radius R, where the latter is defined so that when the 3-measure $d\Omega_3 = d\Omega_3(k)$ reduces to the 2-measure defined above, then $2\pi R$ is the circumference. The Kretschmann scalar for the metric (3.22) is $72K^2/R^8$, where the 5D curvature K depends on the mass M at the centre of the 3-geometry. This scalar is the same as Eq. (3.12) for the cosmological manifold (3.11). This coincidence and other properties suggest to us that Eqs. (3.11) and (3.22) are geometrically equivalent descriptions of the same situation in different coordinates, or isometries.

To prove this, we need to show a coordinate transformation which takes us from Eq. (3.22) to (3.11) or the reverse. We proceed to give the result as derived by Seahra and Wesson [31], noting that it may be confirmed by computer. The radial transformation is specified by

$$R^2 = (\mu^2 + k)\ell^2 + 2v\ell + (v^2 + K)(\mu^2 + k)^{-1} \quad , \tag{3.23}$$

where μ and v are the functions of t in Eq. (3.11). This is an unusual mapping, which may repay further investigation. The corresponding temporal transformation turns out to have different forms, depending on whether $k = 0$ or $k = \pm 1$. To present these forms, we introduce a dummy variable $u = u(t)$ and the function $v = v(t, \ell) \equiv k[(\mu^2 + k)\ell + v](kK)^{-1/2}\mu^{-1}$. Then for $k = 0, \pm 1$ we have respectively:

$$T = \frac{1}{K}\int \left[\frac{v^2}{\mu^3}v' - \frac{v(v^2+K)}{\mu^4}\mu'\right] du + \frac{1}{K}\left[\frac{\mu^3\ell^3}{3} + \mu v\ell^2 + \left(\frac{v^2+K}{\mu}\right)\ell\right] \quad ,$$

$$\tag{3.24}$$

$$T = \frac{1}{k}\int \left[\frac{v'}{\mu} - \left(\frac{v}{\mu^2+k}\right)\mu'\right] du + \frac{1}{k}\left[\mu\ell - \frac{K}{2(kK)^{1/2}}\ln\left(\frac{1+v}{1-v}\right)\right] \quad .$$

In these equations, it is to be understood that the integrals are over t and that the integrands involve $v = v(u)$ and $\mu = \mu(u)$ with $v' \equiv dv/du$ and $\mu' \equiv d\mu/du$. It should also be noted that while Eqs. (3.24) relate time near the black hole defined by Eq. (3.22) to time in the universe defined by Eq. (3.11), there is a special case of Eq. (3.22) where $kK < 0$ and $k = \pm 1$. However, for this case there is no Killing-vector defined horizon, so this would correspond to a naked singularity, with a negative mass.

We therefore bypass this special case as unphysical, and conclude that Eqs. (3.23) and (3.24) are in general the coordinate transformations that take the metric (3.22) for a 5D topological black hole back to the metric (3.11) for a 5D FLRW-like cosmology.

The isometry just demonstrated invites further analysis based on a comparison with the usual 4D Schwarzschild solution. The latter is commonly presented in coordinates where there is a horizon (defined by mass M) which splits the manifold into parts, the distinction being somewhat artificial from the geometrical viewpoint. This problem is frequently addressed by introducing Kruskal-Szekeres (KS) coordinates, which effectively remove the horizon and extend the geometry. We are naturally interested in seeing if this is possible for the metric constructed from Eqs. (3.23) and (3.24). There are many choices for the parameters involved in these relations, so let us focus on the case where $k = +1$ and $K > 0$. This means that the 3D submanifold is spherical, so we have an ordinary as opposed to a topological black hole. Then it may be shown that for our case there are KS-type coordinates U, V which are related to the R, T coordinates in Eqs. (3.23) and (3.24) by

$$U = \mp[h(\mathrm{R})]e^{R/M}e^{-T/M}|(R - M)/(R + M)|^{1/2} \ ,$$

$$V = \pm e^{R/M}e^{T/M}|(R - M)/(R + M)|^{1/2} \ . \tag{3.25}$$

In these coordinates, the metric for the black hole is

$$dS^2 = M^2(1 + M^2/R^2)e^{-2R/M}dUdV - R^2d\Omega_3^2 \ . \tag{3.26}$$

A detailed investigation of the extended geometry defined by Eq. (3.26), including Penrose-Carter diagrams, appears in Ref. [31]. This clarifies the nature in which the manifold is covered by the coordinates of the cosmological metric (3.11), to which we now return.

According to Eq. (3.11), an observer unaware of the fifth dimension or confined to a hypersurface in it would experience a 4D universe with line element

$$ds^2 = \left(\frac{1}{\mu}\frac{\partial A}{\partial t}\right)^2 dt^2 - A^2\left[\frac{dr^2}{1 - kr^2} + r^2(d\theta^2\sin^2\theta d\phi^2)\right] \ . \tag{3.27}$$

In Section 3.4 above, we outlined several interpretations of this metric, which follow from choosing the two arbitrary functions $\mu(t)$ and $v(t)$ and evaluating its associated matter. We noted that the main feature of the class of solutions (3.11) is a bounce, where the spatial scale factor of Eq. (3.27) goes through a minimum. Now that we know that the cosmological metric is isometric to a black hole, it is easier to see what is involved: The scale factor $A(t, \ell)$ when it passes through the minimum ($\partial A/\partial t = 0$) induces a singularity in the metric which is of the same type as with a conventional black hole ($g_{00} = 0$). But we argued before that this singularity is not geometrical, and indeed it is now clear that it is of the kind found at the event horizon of a black hole. However, an observer in the 4D manifold (3.27) would interpret $A(t)$ as the standard scale factor of an FLRW model if $A(t, \ell)$ evolves to be independent of the fifth dimension. He would then wrongly assume that the universe starts from a big bang in the state with $A = 0$, whereas it actually evolves from a state with $\partial A/\partial t = 0$ which is a big bounce. In the case where the bounce is associated with matter production, as we also mentioned in Section 3.4, it is useful to introduce the mass M of the fluid out to radius r as it is defined by the density ρ, pressure p and the metric (3.27). This is given by the Misner-Sharp-Podurets mass function which was introduced in Section 1.4 [34-36]. For the uniform fluid described by Eq. (3.27), the relevant relations are

$$M = 4\pi A r^3 (\mu^2 + k) = (4\pi/3) r^3 A^3 \rho \ ,$$

$$\partial M/\partial t = -4\pi r^3 A^2 (\partial A/\partial t) p \ .$$

(3.28)

These allow of matter production ($\partial M/\partial t > 0$) both before the bounce ($\partial A/\partial t < 0, p > 0$) and after it ($\partial A/\partial t > 0, p < 0$), at least on the basis of classical theory. However, a proper investigation of this would require quantum theory, which would also help clarify the status of other issues with these models, such as inflation.

In this section, we have suggested that an observer living in a universe with the 4D line element (3.27) might be unaware that it is part of a 5D model of the form (3.11), which is geometrically equivalent to the 5D black hole (3.22). The argument has been mainly mathematical in nature. From a conceptual viewpoint, the idea that the universe is a higher-dimensional black hole may be harder to accept.

3.6. Anisotropic 5D Universes

From the preceding sections, it appears that our desire to find 5D analogs of the standard 4D cosmologies leads to models whose mathematics are easier to understand than their physics. However, several workers have pointed out that the present uniform universe may have evolved from a non-uniform initial state, and that the matter in it could have been produced by some gravitational process just after the big bang. A class of 4D solutions was found by Kasner [37] which has these properties [38] and is unusual even by 4D standards.

The Kasner solutions are typified by having three spatial axes which are each homogeneous but different from each other, so the ordinary space is anisotropic. The machinations of these axes shortly after the big bang has been much studied, particularly in regard to particle production and the smoothing of any initial density perturbations, a phenomenon which has led to the name Mixmaster cosmology. But while a kitchen appliance can mix lumpy ingredients to form a smooth medium, the Kasner-Mixmaster model for early times shows properties "reminiscent of an elastic solid" [38]. In view of its intriguing properties, it is natural to ask if there is a 5D analog of the Kasner class of solutions.

The answer to this is positive, as shown by Ponce de Leon and Wesson [39]. However, the algebra is extensive and tedious: the 5D field equations $R_{AB} = 0$ have to be decomposed into their diagonal and non-diagonal components and solved simultaneously, a process which proves to require the imposition of multiple sets of constraints on the constants that control the evolution of the spatial scale factors. These authors actually found two classes of 5D Kasner-like solutions, one in which the metric coefficients depend only on the time $x^0 = t$, and one in which they depend only on the extra coordinate $x^4 = \ell$. These may be combined to give the general class of solutions:

$$dS^2 = A\ell^{2\beta_0}dt^2 - B\ell^{2\beta_1}t^{2\alpha_1}dx^2 - C\ell^{2\beta_2}t^{2\alpha_2}dy^2 - D\ell^{2\beta_3}t^{2\alpha_3}dz^2 + \varepsilon Et^{2\alpha_4}d\ell^2 \ . \tag{3.29}$$

Here the constants $A-E$ are arbitrary. The parameters $\alpha_{1,2,3,4}$ and $\beta_{0,1,2,3}$ satisfy sets of constraints, as follows:

$$\alpha_1 + \alpha_2 + \alpha_3 + \alpha_4 = 1 \quad , \quad \alpha_1^2 + \alpha_2^2 + \alpha_3^2 + \alpha_4^2 = 1 \quad ,$$

$$\beta_0 + \beta_1 + \beta_2 + \beta_3 = 1 \quad , \quad \beta_0^2 + \beta_1^2 + \beta_2^2 + \beta_3^2 = 1 \quad ,$$

$$\alpha_1\beta_1 + \alpha_2\beta_2 + \alpha_3\beta_3 - \beta_0(\alpha_1 + \alpha_2 + \alpha_3) - \\ \alpha_4(\beta_1 + \beta_2 + \beta_3) = 0 \quad .$$

(3.30)

The class of 5D solutions given by these relations reduces in the appropriate limit to the corresponding 4D solutions of Bianchi-I type. These are anisotropic, but again reduce in the appropriate limit to the homogeneous and isotropic FLRW models of standard cosmology discussed in Chapter 1. The 5D analogs of the 3D flat FLRW models were treated in section 3.2, where it was noted that they obey not only $R_{AB} = 0$ but also $R_{AB} = 0$, meaning that they are 5D-flat. By contrast, the solutions (3.29) above have in general $R_{AB} \neq 0$.

The properties of matter associated with the metric (3.29) can be derived from the induced energy-momentum tensor, following the standard procedure in STM theory detailed in Chapter 2. There results

$$8\pi T_0^0 = \frac{\alpha_1(\alpha_2+\alpha_3)+\alpha_2\alpha_3}{A\ell^{2\beta_0}t^2} \quad ,$$

$$8\pi T_1^1 = \frac{\alpha_2^2+\alpha_3^2+\alpha_2(\alpha_3-1)-\alpha_3}{A\ell^{2\beta_0}t^2} \quad ,$$

$$8\pi T_2^2 = \frac{\alpha_1^2+\alpha_3^2+\alpha_1(\alpha_3-1)-\alpha_3}{A\ell^{2\beta_0}t^2} \quad ,$$

(3.31)

$$8\pi T_3^3 = \frac{\alpha_1^2+\alpha_2^2+\alpha_1(\alpha_2-1)-\alpha_2}{A\ell^{2\beta_0}t^2} \quad .$$

Here the parameters are constrained by Eqs. (3.30). We note that the last three components are invariant under the interchanges $1\leftrightarrow2$, $1\leftrightarrow3$, $2\leftrightarrow3$, a property that they inherit from the field equations. Another property of Eqs. (3.31) is that, for a general choice of parameters, the components satisfy $T = T_0^0 + T_1^1 + T_2^2 + T_3^3 = 0$. This in 4D physics is usually taken to mean that the equation of state is that of radiation (i.e., photons with zero rest mass) or ultra-relativistic matter (i.e., particles with finite rest mass moving close to the speed of light). However, in STM theory, it is well known that when the 4D part of the 5D metric is independent of x^4, the effective energy-momentum tensor has $T = 0$. Our result implies that the reverse implication need not hold. Notwithstanding this, if the early universe was Kasner-like then it was hot.

3.7. Divergent Dark Energy

Cosmology is subject to upheavals, which come from the theoretical side or the observational side. The previous sections in this chapter have examined primarily theoretical alternatives to the conventional big bang. However, there is a convincing group of observational data, largely to do with Type Ia supernovae, which indicates that the expansion of the present universe is accelerating, implying that it is dominated by dark energy (Chapter 1). There is also a significant amount of dark matter, including a small fraction of dark but otherwise ordinary baryons; but overall it appears that luminous matter, in the form of stars and galaxies, is a mere smattering on top of a sea of dark energy. The latter could be simply the cosmological constant of Einstein's theory, whose equation of state ($p = -\rho$) means that it acts like a repulsive fluid; but it could also be a scalar field with similar properties. This fits with the idea that the universe began in an inflationary mode, perhaps as the result of a quantum event.

STM theory has a ready, if classical, explanation for this, based on 5D canonical space. We have already seen one application of this in the bounce metric (3.11), and a detailed discussion of the canonical space occupies Chapter 7. Since all 4D metrics with a cosmological constant Λ can be embedded in 5D canonical space, it is natural to use the latter to model the supernova data. Specifically we will examine a model where Λ is in fact variable, and as a measure of the density of dark energy actually diverges at the big bang.

That the cosmological 'constant' can in fact depend on the extra coordinate so $\Lambda = \Lambda(\ell)$ was found by Mashhoon and Wesson [40], and the astrophysical implications of this were studied by Overduin and others [41]. It is well known that the standard canonical metric gives back the Einstein cosmological constant as $\Lambda = 3/L^2$ where L is a constant length which appears in the embedding (see below). But Λ becomes variable when the extra coordinate is shifted along its own axis by a constant, $\ell \to (\ell - \ell_0)$. Analysis of the field equations $R_{AB} = 0$ shows that there is a class of cosmological models with the shifted canonical metric given by

$$dS^2 = \left(\tfrac{\ell - \ell_0}{L}\right)^2 f(x^\gamma) \eta_{\alpha\beta} dx^\alpha dx^\beta - d\ell^2 \ . \tag{3.32}$$

Here $f(x^\gamma)$ is an arbitrary function of the spacetime coordinates and $\eta_{\alpha\beta} = \mathrm{diag}(1, -1, -1, -1)$ is the metric tensor of Minkowski spacetime. Metrics of the type (3.32) are therefore 4D conformally flat. It is well known that the de Sitter model is of this type. That model has zero 3D curvature and an exponential scale factor with a timescale set by the cosmological constant, properties which are often taken as the basis for inflation. However, other 5D metrics are possible, defining Einstein spaces where the 4D Ricci tensor is equal to Λ times the metric tensor. We find that Eq. (3.32) is of this type, with an effective cosmological constant

$$\Lambda = \frac{3}{L^2} \left(\frac{\ell}{\ell - \ell_0}\right)^2 \ . \tag{3.33}$$

This expression reduces to the Einstein value $\Lambda = 3/L^2$ when $\ell_0 \to 0$, but in general Λ is variable in spacetime because $\ell = \ell(s)$. Also, when $\ell \to \ell_0$ the cosmological 'constant' diverges. In this regard, Eq. (3.33) also represents a formal connection between STM theory and M theory with a singular hypersurface.

To obtain Λ as an explicit function of 4D proper time s requires in general a solution for $\ell = \ell(s)$ from the 5D geodesic equation (Section 7.3). This also shows that there is an extra force (per unit mass) F^μ which acts in spacetime as a result of the fifth dimension. A particularly simple and generic solution for $\ell(s)$ follows from the assumption of null 5D geodesics with $dS^2 = 0$. This is a logical assumption, and with an appropriate choice of boundary condition and sign we write this solution as $\ell = \ell_0 \exp(s/L)$. Then we can evaluate the cosmological 'constant' as

$$\Lambda = \frac{3}{L^2} \frac{1}{\left(1-e^{-s/L}\right)^2} \ .$$

(3.34)

This is unbounded at the big bang ($s = 0$), and decays asymptotically to $3/L^2$ (as $s \to \infty$). We can also evaluate the fifth force noted above, which becomes:

$$F^\mu = -\frac{1}{L} \frac{dx^\mu}{ds} \frac{1}{\left(e^{s/L}-1\right)} \ .$$

(3.35)

This is also unbounded at the big bang ($s = 0$), but decays asymptotically to zero ($s \to \infty$). It represents an acceleration which is always toward a local origin for particles dispersing as the universe expands ($F^\mu < 0$ for $dx^\mu/ds > 0$). That is to say, it assists gravity.

The astrophysical consequences of the model outlined above have been examined in detail [41]. It is inherently inflationary at early times, but with a cosmological 'constant' which persists at late times. This is positive in sign, so galaxies with separation r feel the usual force (per unit mass) $\Lambda r/3$ pushing them apart. However, the fifth force defined by Eq. (3.35) acts in the opposite direction, especially at early times, so compared to the conventional models galaxy formation should have been easier. Additional supernova data might help to discriminate between the 5D and 4D approaches, especially at redshifts in the range $1 < z < 3$.

3.8. Conclusion

In the preceding five sections we examined ways in which the initial singularity in 4D may if so desired be replaced by a more physical event in 5D. It could be that the big bang is an artefact due to an unfortunate choice of coordinates in a flat 5D universe. Or it could be a shock wave, also in a flat 5D manifold. Alternatively, if the canonical metric has significant dependence on the extra coordinate, there can be a bang and a bounce, the latter coming later and its divergent matter properties being responsible for the physics we conventionally attribute to the fireball. (Other consequences of the canonical metric will be investigated in Chapter 7.) Another possibility – unusual but consistent with the geometry – is that our 4D universe as 'seen' from 5D may resemble a black hole. Then there is an extension of the Kasner solutions from 4D to 5D which implies that the early universe may not have been regular but rather anisotropic and chaotic. And lastly, there is a version of canonical space where the cosmological 'constant' is variable, measuring the density of a kind of dark energy which diverges at time zero.

The options outlined above are not exhaustive, and there could be other viable alternatives to the conventional big bang. However, even five alternatives to the standard singularity might be considered by some workers to be too rich a field of possibilities. A strict application of Ockham's razor would seem to imply that we should revert to the unique and relatively simple concept of a big bang. The real issue here, though, is something else: If there are only four dimensions to the world then it is close to inevitable that there was a big bang, whereas the existence of one (or more) extra dimension implies an alternative type of event.

References

[1] Hoyle, F., *Astrophys. J.* **196**, 661 (1975).
[2] Vilenkin, A., *Phys. Lett.* **B117**, 25 (1982).
[3] Bonnor, W.B., *J. Math. Mech.* **9**, 439 (1960); Wesson, P.S., *Astron. Astrophys.* **151**, 276 (1985).
[4] Ponce de Leon, J., *Gen. Rel. Grav.* **20**, 539 (1988).
[5] Wesson, P.S. and Seahra, S.S., *Astrophys. J.* **558**, L75 (2001); Seahra, S.S. and Wesson, P.S., *Class. Quant. Grav.* **19**, 1139 (2002).
[6] Wesson, P.S., *Space-Time-Matter*, 2nd. ed. (World Scientific, Singapore, 2007).
[7] Overduin, J.M., *Astrophys. J.* **517**, L1 (1999).
[8] Rindler, W., *Relativity: Special, General and Cosmological* (Oxford Un. Press, Oxford, 2001).
[9] Linde, A.D., *Inflation and Quantum Cosmology* (Academic, Boston, 1990); *Phys. Scripta* **T36**, 30 (1991); *Sci. Am.* **271** (11), 48 (1994).
[10] Wesson, P.S., Liu, H. and Seahra, S.S., *Astron. Astrophys.* **358**, 425 (2000).
[11] Ponce de Leon, J., *Int. J. Mod. Phys.* **D12**, 1053 (2003).
[12] Abolghasem, G., Coley, A.A. and McManus, D.J., *J. Math. Phys.* **37**, 361 (1996).
[13] Liebbrandt, G., *Non-Covariant Gauges* (World Scientific, Singapore, 1994).
[14] Leonard, S., Lake, K., *Astrophys. J.* **441**, L55 (1995).
[15] Perlmutter, S., *Phys. Today* **56** (4), 53 (2003).
[16] Overduin, J.M. and Wesson, P.S., *The Light / Dark Universe* (World Scientific, Singapore, 2008).
[17] Mashhoon, B., Liu, B. and Wesson, P.S., *Phys. Lett.* **B331**, 305 (1994).

[18] Liu, H. and Mashhoon, B., *Ann. Phys.* **4**, 565 (1995).

[19] Liu, H. and Mashhoon, B., *Phys. Lett.* **A272**, 26 (2000).

[20] Liu, H. and Wesson, P.S., *Astrophys. J.* **562**, 1 (2001).

[21] Chang, B., Liu, H. and Xu, L., *Mod. Phys. Lett.* **A20**, 923 (2005).

[22] Liu, H. *et al.*, *Mod. Phys. Lett.* **A20**, 1973 (2005).

[23] Zhang, C., Liu, H. and Xu, L., Wesson, P.S., *Mod. Phys. Lett.* **A21**, 571 (2006).

[24] Liu, M., Liu, H., Xu, L. and Wesson, P.S., *Mod. Phys. Lett.* **A21**, 2937 (2006).

[25] Liu, H., *Phys. Lett.* **B560**, 149 (2003).

[26] Xu, L., Liu, H. and Wang, B., *Chin. Phys. Lett.* **20**, 995 (2003).

[27] Wang, B., Liu, H. and Xu, L., *Mod. Phys. Lett.* **A19**, 449 (2004).

[28] Ponce de Leon, J., *Mod. Phys. Lett.* **A17**, 2425 (2002).

[29] Liko, T. and Wesson, P.S., *Int. J. Mod. Phys.* **A20**, 2037 (2005).

[30] Wesson, P.S., *Five-Dimensional Physics* (World Scientific, Singapore, 2006), 53.

[31] Seahra, S.S. and Wesson, P.S., *J. Math. Phys.* **44**, 5664 (2003).

[32] Seahra, S.S. and Wesson, P.S., *Gen. Rel. Grav.*, **37**, 1339 (2005).

[33] Fukui, T., Seahra, S.S. and Wesson, P.S., *J. Math. Phys.* **42**, 5195 (2001).

[34] Misner, C.W. and Sharp, D.H., *Phys. Rev.* **B136**, 571 (1964).

[35] Podurets, M.A., *Sov. Astron. (A.J.)* **8**, 19 (1964).

[36] Henriksen, R.N., Emslie, A.G. and Wesson, P.S., *Phys. Rev.* **D27**, 1219 (1983); Wesson, P.S., *Phys. Rev.* **D34**, 3925 (1986).

[37] Kasner, E., *Ann. J. Math.* **43**, 217 (1921).

[38] Misner, C.W., Thorne, K.S. and Wheeler, J. A., *Gravitation* (Freeman, San Francisco, 1973), 806.

[39] Ponce de Leon, J. and Wesson, P.S., *Europ. Phys. Lett.* **84**, 20007 (2008).

[40] Mashhoon, B. and Wesson, P.S., *Class. Quant. Grav.* **21**, 3611 (2004); *Gen. Rel. Grav.* **39**, 1403 (2007); Ponce de Leon, J., arXiv:0712.4301 (2007).

[41] Overduin, J.M., Wesson, P.S., Mashhoon, B., *Astron. Astrophys.* **473**, 727 (2007); Wesson, P.S., *Astron. Astrophys.* **441**, 41 (2005).

Chapter 4

Astrophysics in 5D

4.1. Introduction

Since Space-Time-Matter theory contains general relativity, many astrophysical systems have the same status under the one as the other. However, the wider scope of 5D theory as compared to 4D means that we can find new solutions of physical relevance, and it is mainly these which we will look at in the present chapter. Astrophysical systems are generally described by metrics with less symmetry than those of cosmology – typically spherically symmetric in ordinary space as opposed to homogeneous and isotropic – and the more restricted nature of the geometry leads to more constrained physics than the free-for-all of the preceding chapter.

We will in fact begin our studies with the case of 3D spherical symmetry and apply it to an isothermal cloud of matter (Section 4.2), followed by a self-similar ball of radiation (Section 4.3). Then we will briefly consider systems with axial and cylindrical symmetry, which may be relevant to the early universe (Section 4.3). It is, of course, widely believed that the early universe was inflationary and dominated by vacuum energy. In such an environment, bubbles may have been the precursors of galaxies and other structures, so we will look at shell-like systems and related metrics which are 5D-flat (Section 4.5). Lastly, we will note that in 5D the de Sitter metric with a negative cosmological constant can support periodic perturbations or vacuum waves (Section 4.6). The latter solution may be of marginal relevance to astrophysics, but is of significance because it forms the basis of subsequent work on canonical space and the cosmological 'constant,' as well as the applications of the 5D vacuum to particles and waves (Chapters 7 and 8). As usual, the last section is a summary, where we will also outline implications of 5D for other astrophysical systems.

4.2. Systems with 3D Spherical Symmetry

It is common practice to model astrophysical systems which are not grossly asymmetric as spherical clouds of matter with inhomogeneous density profiles. For example, rich or Abell clusters of galaxies can be modeled as static, isothermal spheres with power-law density profiles. In this section, we will study a general metric and then give a static solution with inverse-square isothermal matter. The latter is the 5D analog of a well-known 4D solution [2]. There are presumably many solutions of the 5D equations we will present below which contain known 4D solutions.

We take a spherically-symmetric metric in the form

$$dS^2 = e^\nu dt^2 - e^\lambda dr^2 - R^2 d\Omega^2 + \varepsilon\, e^\mu d\ell^2 \ . \tag{4.1}$$

Here the metric coefficients ν, λ, R, μ can depend in general on the time t, radius r and the extra coordinate ℓ. The nonzero components of R_{AB} for Eq. (4.1) are

$$
\begin{aligned}
R_{00} = &-\frac{\ddot{\lambda}}{2} - \frac{\ddot{\mu}}{2} - 2\frac{\ddot{R}}{R} + \frac{\dot\nu\dot\lambda}{4} + \frac{\dot\nu\dot\mu}{4} + \frac{\dot\nu\dot R}{R} - \frac{\dot\lambda^2}{4} - \frac{\dot\mu^2}{4} \\
&+ e^{\nu-\lambda}\left(\frac{\nu''}{2} + \frac{\nu'^2}{4} - \frac{\nu'\lambda'}{4} + \frac{\nu'\mu'}{4} + \frac{\nu'R'}{R}\right) \\
&+ \varepsilon e^{\nu-\mu}\left(-\frac{\overset{**}{\nu}}{2} - \frac{\overset{*}{\nu}{}^2}{4} + \frac{\overset{*}{\nu}\overset{*}{\mu}}{4} - \frac{\overset{*}{\nu}\overset{*}{\lambda}}{4} - \frac{\overset{*}{\nu}\overset{*}{R}}{R}\right) \ ,
\end{aligned}
$$

$$
R_{01} = -\frac{\dot\mu'}{2} - \frac{\dot\mu\mu'}{4} + \frac{\nu'\dot\mu}{4} + \frac{\dot\lambda\mu'}{4} + \frac{\dot\lambda R'}{R} + \frac{\nu'\dot R}{R} - \frac{2\dot R'}{R} \ ,
$$

$$R_{04} = -\frac{\overset{*}{\dot\lambda}}{2} - \frac{\dot\lambda\overset{*}{\lambda}}{4} + \frac{\dot\lambda\overset{*}{\nu}}{4} + \frac{\overset{*}{\lambda}\dot\mu}{4} + \frac{\dot\mu\overset{*}{R}}{R} + \frac{\overset{*}{\nu}\dot R}{R} - \frac{2\overset{*}{\dot R}}{R} \ ,$$

$$R_{11} = -\frac{\nu''}{2} - \frac{\mu''}{2} - \frac{\nu'^2}{4} - \frac{\mu'^2}{4} + \frac{\lambda'\nu'}{4} + \frac{\lambda'\mu'}{4} + \frac{\lambda'R'}{4}$$
$$- \frac{2R''}{R} + e^{\lambda-\nu}\left(\frac{\ddot\lambda}{2} + \frac{\dot\lambda^2}{4} - \frac{\dot\lambda\dot\nu}{4} + \frac{\dot\lambda\dot\mu}{4} + \frac{\dot\lambda\dot R}{R}\right)$$
$$+ \varepsilon e^{\lambda-\mu}\left(\frac{\overset{**}{\lambda}}{2} + \frac{\overset{*}{\lambda}{}^2}{4} + \frac{\overset{*}{\lambda}\overset{*}{\nu}}{4} - \frac{\overset{*}{\lambda}\overset{*}{\mu}}{4} + \frac{\overset{*}{\lambda}\overset{*}{R}}{R}\right) \ ,$$

$$(4.2)$$

$$R_{14} = -\frac{\overset{*}{\nu'}}{2} - \frac{\nu'\overset{*}{\nu}}{4} + \frac{\overset{*}{\lambda}\nu'}{4} + \frac{\mu'\overset{*}{\nu}}{4} + \frac{\overset{*}{\lambda}R'}{R} + \frac{\mu'\overset{*}{R}}{R} - \frac{2\overset{*}{R'}}{R} \ ,$$

$$R_{22} = 1 + R^2 e^{-\nu}\left[\frac{\dot R^2}{R^2} + \frac{\ddot R}{R} - \frac{\dot R}{2R}(\dot\nu - \dot\lambda - \dot\mu)\right]$$
$$- R^2 e^{-\lambda}\left[\frac{R'^2}{R^2} + \frac{R''}{R} + \frac{R'}{2R}(\nu' - \lambda' + \mu')\right]$$
$$+ \varepsilon R^2 e^{-\mu}\left[\frac{\overset{*}{R}{}^2}{R^2} + \frac{\overset{**}{R}}{R} + \frac{\overset{*}{R}}{2R}(\overset{*}{\nu} + \overset{*}{\lambda} - \overset{*}{\mu})\right] \ ,$$

$$R_{33} = R_{22}\sin^2\theta \ ,$$

$$R_{44} = -\frac{\overset{**}{\nu}}{2} - \frac{\overset{*}{\nu}{}^2}{4} - \frac{\overset{**}{\lambda}}{2} - \frac{\overset{*}{\lambda}{}^2}{4} + \frac{\overset{*}{\mu}\overset{*}{\nu}}{4} + \frac{\overset{*}{\mu}\overset{*}{\lambda}}{4} + \frac{\overset{*}{\mu}\overset{*}{R}}{R} - \frac{2\overset{**}{R}}{R}$$
$$- \varepsilon e^{\mu-\nu}\left(\frac{\ddot\mu}{2} + \frac{\dot\mu^2}{4} - \frac{\dot\mu\dot\nu}{4} + \frac{\dot\mu\dot\lambda}{4} + \frac{\dot\mu\dot R}{R}\right)$$
$$+ \varepsilon e^{\mu-\lambda}\left(\frac{\mu''}{2} + \frac{\mu'^2}{4} + \frac{\mu'\nu'}{4} - \frac{\mu'\lambda'}{4} + \frac{\mu'R'}{R}\right) \ .$$

Here as before, a dot denotes $\partial/\partial t$, a prime denotes $\partial/\partial r$ and an asterisk denotes $\partial/\partial\ell$. The nonzero components of T_β^α for Eq. (4.1) are

$$8\pi T_0^0 = -e^{-\nu}\left(\frac{\dot\lambda\dot\mu}{4} + \frac{\dot R\dot\mu}{R}\right) + e^{-\lambda}\left(\frac{R'\mu'}{R} - \frac{\lambda'\mu'}{4} + \frac{\mu''}{2} + \frac{\mu'^2}{4}\right)$$
$$- \varepsilon e^{-\mu}\left(\frac{\overset{**}{\lambda}}{2} + \frac{\overset{*}{\lambda}{}^2}{4} - \frac{\overset{*}{\mu}\overset{*}{\lambda}}{4} + \frac{\overset{*}{R}\overset{*}{\lambda}}{R} - \frac{\overset{*}{R}\overset{*}{\mu}}{R} + \frac{\overset{*}{R}{}^2}{R^2} + \frac{2\overset{**}{R}}{R}\right) \ ,$$

$$8\pi T_0^1 = -e^{-\lambda}\left(\frac{\dot\mu'}{2} + \frac{\dot\mu\mu'}{4} - \frac{\nu'\dot\mu}{4} - \frac{\dot\lambda\mu'}{4}\right) \ ,$$

$$(4.3)$$

$$8\pi T_1^1 = -e^{-\nu}\left(\frac{\ddot{\mu}}{2} + \frac{\dot{\mu}^2}{4} - \frac{\dot{\nu}\dot{\mu}}{4} + \frac{\dot{R}\dot{\mu}}{R}\right) + e^{-\lambda}\left(\frac{R'\mu'}{R} + \frac{\nu'\mu'}{4}\right)$$
$$- \varepsilon e^{-\mu}\left(\frac{\dot{R}^2}{R^2} + \frac{2\ddot{R}}{R} + \frac{\dot{R}\dot{\nu}}{R} - \frac{\ddot{R}\dot{\mu}}{R} + \frac{\ddot{\nu}}{2} + \frac{\dot{\nu}^2}{4} - \frac{\dot{\nu}\dot{\mu}}{4}\right) ,$$

$$8\pi T_2^2 = -e^{-\nu}\left(\frac{\dot{R}\dot{\mu}}{2R} - \frac{\dot{\nu}\dot{\mu}}{4} + \frac{\dot{\lambda}\dot{\mu}}{4} + \frac{\ddot{\mu}}{2} + \frac{\dot{\mu}^2}{4}\right)$$
$$+ e^{-\lambda}\left(\frac{R'\mu'}{2R} + \frac{\mu''}{2} + \frac{\mu'^2}{4} - \frac{\lambda'\mu'}{4} + \frac{\nu'\mu'}{4}\right)$$
$$- \varepsilon e^{-\mu}\left(\frac{\ddot{R}}{R} + \frac{\dot{R}\dot{\nu}}{2R} + \frac{\dot{R}\dot{\lambda}}{2R} - \frac{\ddot{R}\dot{\mu}}{2R} + \frac{\ddot{\nu}}{2}\right.$$
$$\left. + \frac{\dot{\nu}^2}{4} + \frac{\ddot{\lambda}}{2} + \frac{\dot{\lambda}^2}{4} + \frac{\dot{\nu}\dot{\lambda}}{4} - \frac{\dot{\nu}\dot{\mu}}{4} - \frac{\dot{\mu}\dot{\lambda}}{4}\right) ,$$

$$T_3^3 = T_2^2 .$$

We see that in general the fluid is anisotropic ($T_1^1 \neq T_2^2$), and that in the time-dependent case there is a radial flow of energy ($T_0^1 \neq 0$). However, it is apparent that previous relations simplify greatly for the time-independent case, to which we now turn.

A class of exact solutions to $R_{AB} = 0$ that is static is given by

$$dS^2 = \left(\frac{r}{r_0}\right)^{2(\alpha+1)} \ell^{2(\alpha+3)/\alpha} dt^2 - (3-\alpha^2)\ell^2 dr^2$$
$$- \ell^2 r^2 d\Omega^2 + 3(3\alpha^{-2} - 1)r^2 d\ell^2 . \qquad (4.4)$$

Here r_0 is a constant and α is a parameter related to the properties of matter. The latter can be obtained by substituting the metric (4.4) into the energy-momentum tensor (4.3) to give

$$8\pi T_0^0 = \frac{2-\alpha^2}{(3-\alpha^2)\ell^2 r^2} ,$$

$$8\pi T_1^1 = -\frac{\alpha^2+2\alpha}{(3-\alpha^2)\ell^2 r^2} ,$$

$$8\pi T_2^2 = -\frac{\alpha^2+2\alpha+1}{(3-\alpha^2)\ell^2 r^2} , \qquad (4.5)$$

$$T_3^3 = T_2^2 \; .$$

If we use $\overline{p} \equiv -(T_1^1 + T_2^2 + T_3^3)/3$, then we have

$$8\pi\overline{p} = \frac{\alpha^2 + 2\alpha + 2/3}{(3-\alpha^2)\ell^2 r^2} \;, \quad 8\pi\rho = \frac{2-\alpha^2}{(3-\alpha^2)\ell^2 r^2} \;, \tag{4.6}$$

so the equation of state is isothermal. The physical properties of this solution have been explored by Billyard and Wesson [1]. They have also solved the 5D geodesic equation for a particle moving in the field of Eq. (4.4), and have pointed out how the latter can be applied to astrophysical problems such as the gravitational lensing of QSOs by clusters of galaxies. The main mathematical property of the solution is that the 5D Kretschmann scalar is $K \equiv R_{ABCD}R^{ABCD} = 16/3r^4\ell^4$. It is therefore not flat in 5D, like some other solutions to be examined later in this chapter.

4.3. Balls of Radiation

Self-similarity is a useful technique, well suited to modelling astrophysical systems which typically have no well-defined boundaries. It consists basically in using the absence of physical scales to define a single, dimensionless variable, and rewriting the equations in terms of this variable so that they are easier to solve than the original forms. The technique has been extensively applied to problems in general relativity (see Section 1.4). In the present section, we wish to apply self-similarity to the 3D spherically-symmetric situation considered above. Our aim is to illustrate the technique by applying it to a relatively simple system, namely a centrally-condensed fluid with the equation of state of hot matter. We are particularly interested in deriving the morphology and thermo-dynamics of such a 'ball' of radiation.

Before proceeding, it is well to remark that self-similarity is not a covariant technique, meaning it may be apparent in one set of coordinates but not in another. An example is provided in 4D by the Milne cosmology, which is derived in some textbooks by applying a coordinate trans-formation to Minkowski space. The Milne metric had a length scale defining the curvature of the 3D subspace, while the Minkowski metric is obviously free of scales. However, the technique is still useful, especially if the 3D part of a metric is close to flat. Suppose the objective is to find the density of a fluid as a function of the cosmic time and the distance from a locally-defined centre. Then in general, given that there are no dimensional scales present, it is possible to write $\rho = (c^2/Gr^2)f(\xi)$, where $\xi \equiv ct/r$ is the dimensionless similarity variable. The existence of this, plus the fundamental constants, means that the governing relations can be written as ordinary differential equations in ξ rather than partial differential equations in t and r. Thus the mathematics becomes considerably simpler, provided a coordinate frame exists in which the physics is free of scales.

Below, we will analyze a solution for an expanding ball of radiation discussed by Ponce de Leon and Wesson [3]. References to other work can be found there; and because the account is rather long, we will concentrate on its most important aspects. Though it may not always be apparent, the similarity variable will be chosen as $\xi = t/r$, and we will absorb the fundamental constants to streamline the working.

We write the metric in the same form as in the preceding section, but express the 44-part of the metric in terms of a scalar field, $g_{44} = \varepsilon\Phi^2$. Here $\varepsilon = +1$ for a timelike extra dimension and $\varepsilon = -1$ for a spacelike one. In order to obtain matter with the equation of state of radiation or ultrarelativistic particles, we assume that all the metric coefficients depend on time t and radius r but not the extra coordinate $x^4 = \ell$. This simplifies the 5D field equations, $R_{AB} = 0 \ (A, B = 0 - 4)$. These are further simplified by the assumption of self-similarity. But for brevity, we omit the details and just state the general metric and the solution:

$$dS^2 = e^\nu dt^2 - e^\lambda dr^2 - R^2 d\Omega^2 + \varepsilon \, \Phi^2 dy^2 \quad , \tag{4.7}$$

$$dS^2 = B^2 \left[\frac{3r^2}{\alpha^2} dt^2 - t^2 dr^2 - \frac{t^2 r^2}{(3-\alpha^2)} d\Omega^2 \right]$$
$$\pm C^2 r^{2(\alpha+1)} t^{2(\alpha+3)/\alpha} dy^2 \quad . \tag{4.8}$$

Here B and C are arbitrary constants, where B is an inverse length and C is a length raised to a power that depends on the other constant in the solution, namely α. This is dimensionless, and is related to the properties of matter, as will be seen below. It is also related to the time evolution of the extra dimension, because Eq. (4.8) implies

$$\frac{1}{\Phi}\frac{d\Phi}{dt} = \frac{\alpha+3}{\alpha t} \quad . \tag{4.9}$$

Since $0 < \alpha^2 < 3$, it follows that either $-\sqrt{3} < \alpha < 0$ or $0 < \alpha < \sqrt{3}$. In both cases $(\alpha + 3)$ is positive. Therefore, for negative (positive) α the size of the extra dimension decreases (increases) monotonically with time.

The components of the induced energy-momentum tensor for the metric (4.8) can be obtained by following the usual STM procedure (Chapter 2). They are

$$T_0^0 = \frac{2}{B^2 t^2 r^2} \quad , \quad T_1^1 = -\frac{2\alpha^2}{3B^2 t^2 r^2} \quad ,$$

$$T_2^2 = T_3^3 = -\frac{3-\alpha^2}{3B^2 t^2 r^2} \quad , \quad T_0^1 = -\frac{2}{B^2 t^3 r} \quad . \tag{4.10}$$

For the physical interpretation of these components, we notice that: (i) for $\alpha^2 \neq 1$ the principal stresses are unequal; (ii) there is a radial flux of energy, $T_0^1 \neq 0$; (iii) the energy conditions are satisfied via

$$T_0^0 > 0 \quad \text{and} \quad T_0^0 > |T_i^i| \quad (i = 1,2,3 \text{ with } \alpha^2 < 3) \; ; \tag{4.11}$$

and (iv) the trace of the energy-momentum tensor vanishes:

$$T_0^0 + T_1^1 + T_2^2 + T_3^3 = 0 \quad . \tag{4.12}$$

This is the usual relation for photons or ultrarelativistic particles, and in terms of the average pressure $\bar{p} = -(T_1^1 + T_2^2 + T_3^3)/3$ and density $\rho = T_0^0$ is commonly written as $\bar{p} = \rho/3$.

The properties of matter given above are notable in having unequal pressures in the radial and azimuthal directions. This situation occurs frequently when new 5D solutions are reduced to their 4D counterparts, and raises a question: Is the cloud of matter static and intrinsically anisotropic as might happen if for example there is a radial field with no orthogonal counterpart; or is the cloud moving, giving the impression of unequal pressures in what is really an isotropic distribution? The second option is possible, because we have not so far inquired about what can be learned from the equation of motion. It transpires that the latter may be avoided, by assuming that the solution (4.8) exists in comoving coordinates, that can be changed to non-comoving ones, under the condition that in the moving frame the pressure is isotropic. The details of this procedure are given elsewhere [3]. It consists, in essence, of using previous relations to evaluate the quantities in the general form of the energy-momentum tensor for anisotropic matter with energy flux. This can be covariantly given as

$$T_{\mu\nu} = (\rho + p_\perp)u_\mu u_\nu - p_\perp g_{\mu\nu} + (p - p_\perp)\chi_\mu\chi_\nu \\ + q_\mu u_\nu + q_\nu u_\mu \ . \tag{4.13}$$

Here u^μ is the four-velocity, χ^μ is a unit spacelike vector orthogonal to u^μ, ρ is the energy density, p is the pressure in the direction of χ_μ, and $p_\perp$ is the pressure in the two-space orthogonal to χ_μ. The energy flux is described by the 4-vector q_μ and obeys $q_\mu u^\mu = 0$. Some tedious algebra shows that Eq. (4.13) can indeed be completely specified.

It is then possible to proceed further, and obtain thermodynamic results for the solution. However, in order to ensure that a model with heat conduction is physically acceptable, besides the usual energy conditions, an appropriate set of thermodynamic relations must be satisfied. These are: (i) baryon conservation

$$\nabla_\mu(nu^\mu) = 0 \ , \tag{4.14}$$

where n is the particle density; (ii) the first law of thermodynamics

$$Td(S/n) = d(\rho/n) + pd(1/n) \ , \tag{4.15}$$

where T is the temperature and S is the entropy density; (iii) the temperature-gradient law

$$q_\mu = -k[(\delta^\nu_\mu - u_\mu n^\nu)\partial_\nu T + T\nabla_\nu u_\mu u^\nu] \ , \tag{4.16}$$

where k is the thermal conductivity ($k \geq 0$); and (iv) positive entropy production

$$\nabla_\mu S^\mu \geq 0 \ , \tag{4.17}$$

where $S^\mu = Su^\mu + q^\mu/T$. It turns out that the model under consideration satisfies all of these conditions. Introducing appropriate constants, it may be shown that the main quantities of interest are given by:

$$n = \frac{n_0}{(B^2 tr)^3} \ , \tag{4.18}$$

$$T = \frac{T_0}{B\sqrt{rt}} \ , \tag{4.19}$$

$$\rho = B^2(3 - \alpha^2)\left(\frac{T}{T_0}\right)^4 \ . \tag{4.20}$$

In addition to these, the thermal conductivity k can be determined, though it has a complicated form. The important point is that the above quantities behave in physically reasonable ways, interpreted in the coordinate frame where the ball of radiation is expanding: the number density (4.18) decreases smoothly with time and distance from the centre; the temperature (4.19) likewise; and the density varies with the temperature (4.20) in accordance with the Stefan-Boltzmann law.

We have paid some attention to deriving the thermodynamic properties of the solution (4.8), because it is a subject often given little notice in classical relativity. The method followed above can be applied to other solutions, and is important because the temperature (for example) is a quantity that can be determined directly from observational data in some cases. That said, there are other properties of the solution which are of a more classical nature and deserve note. Thus, it exists for both a timelike and spacelike extra dimension, unlike the standard cosmologies of Section 3.2 which require $g_{44} < 0$ or the vacuum waves of Section 4.6 which require $g_{44} > 0$. Also, it has $R_{ABCD} \neq 0$, like the solution in the preceding section. In fact, the Kretschmann scalar $K \equiv R_{ABCD}R^{ABCD}$ is proportional to $(rt)^{-4}$ and is therefore singular on the hypersurface $rt = 0$, a property typical of self-similar solutions.

That the solution (4.8) is indeed self-similar may be verified by forming the Lie derivative of the metric, defined by

$$\mathcal{L}_\eta g_{AB} \equiv \partial_C g_{AB} \eta^C + g_{CB} \partial_A \eta^C + g_{AC} \partial_B \eta^C .$$

$$(4.21)$$

In the present case, there is a 5D homothetic Killing vector given by

$$\mathcal{L}_\eta g_{AB} = 2 g_{AB} ,$$

$$(4.22)$$

$$\eta^A = \frac{1}{2} \left[t, r, 0, 0, -\left(\frac{3+\alpha^2}{\alpha} \right) \ell \right] .$$

Consequently the spacetime part of Eq. (4.8) exhibits self-similar symmetry, where all dimensionless quantities are functions of a single independent variable. Specifically

$$\mathcal{L}_\xi g_{\lambda\rho} = 2 g_{\lambda\rho} ,$$

$$(4.23)$$

$$\xi^\mu = \frac{1}{2} (t, r, 0, 0) .$$

The self-similar nature of the spacetime metric becomes evident in the coordinates $\bar{t}, \bar{r}$, which are related to the original ones t, r by

$$t = \sqrt{A} B^{-b} \bar{t}^{(1-b)} \quad , \quad r = \sqrt{A} B^{(b-1)} \bar{r}^b \quad , \quad b \equiv \frac{1}{\sqrt{3-\alpha^2}} \quad , \quad (4.24)$$

where A is a dimensionless constant. In these coordinates the 4D metric becomes

$$ds^2 = \frac{3A^2(1-b)^2}{\alpha^2} \xi^{-2b} d\bar{t}^2$$
$$- \frac{A^2}{(3-\alpha^2)} \xi^{2(1-b)} [d\bar{r}^2 + \bar{r}^2 d\Omega^2] , \qquad (4.25)$$

where the similarity variable is $\xi = (\bar{t}/\bar{r})$. We see that the 5D solution (4.8) is, strictly speaking, only self-similar in its 4D part. Also, the concept depends on the choice of coordinates, as stated near the beginning of this section.

4.4. Systems with Axial and Cylindrical Symmetry

In the preceding sections we studied spherically-symmetric solutions because they are relevant to astrophysics. However, systems with axial and cylindrical symmetry are also relevant, notably in two ways. First, axially-symmetric solutions can describe objects with spin. In this connection, we note that there has been considerable work on 5D Kerr-like solutions [4-10]. However, some of these solutions are contrived and none involves x^4-dependence. Introducing the latter is difficult, and attempts to do so both algebraically and by computer have proven fruitless. This means that the vintage problem of finding a realistic matter source for the external 4D Kerr metric (which is a natural one for the induced-matter approach) remains unsolved. Second, cylindrically-symmetric solutions offer new insight on the question of the inevitability of cosmological singularities. Perhaps unexpectedly, exact solutions of the 4D field equations have been found which while somewhat unphysical have extended matter of cosmological type but no singularities [11-14]. The interest caused by these 4D solutions has led to their generalization to 5D [15, 16]. Following the last-noted reference, we proceed to a brief account of these solutions.

Consider the metric

$$dS^2 = A^2(dt^2 - dr^2) - B^2\,dy^2 - C^2 dz^2 - D^2 d\ell^2 \quad, \quad (4.26)$$

Where the functions A, B, C, D depend only on t and r. As we know, this condition implies an equation of state $T_\alpha^\alpha = 0$, where the components of the energy-momentum tensor are:

$$8\pi T_0^0 = \frac{1}{A^2}\left[\frac{\ddot{D}}{D} - \frac{\dot{A}\dot{D}}{AD} - \frac{A'D'}{AD}\right] \quad,$$

$$8\pi T_0^1 = -\frac{1}{A^2}\left[\frac{\dot{D}'}{D} - \frac{A'\dot{D}}{AD} - \frac{\dot{A}D'}{AD}\right] \quad,$$

$$8\pi T_1^1 = -\frac{1}{A^2}\left[\frac{D''}{D} - \frac{\dot{A}\dot{D}}{AD} - \frac{A'D'}{AD}\right] \quad, \qquad (4.27)$$

$$8\pi T_2^2 = -\frac{1}{A^2}\left[-\frac{\dot{B}\dot{D}}{BD} + \frac{B'D'}{BD}\right] \quad,$$

$$8\pi T_3^3 = -\frac{1}{A^2}\left[-\frac{\dot{C}\dot{D}}{CD} + \frac{C'D'}{CD}\right] \quad.$$

These can be evaluated given a solution for A, B, C, D of the field equations $R_{AB} = 0$, and four cases have been studied in the induced-matter picture by Chatterjee, Wesson and Billyard [16]. Their properties are similar, however, so we quote just one:

$$A = \cosh(2qt)\cosh^4(qr) \quad,$$

$$B = \frac{1}{q}\cosh(2qt)\cosh(qr)\sinh(qr) \quad,$$

$$C = \cosh(2qt)\cosh^2(qr) \quad,$$

$$D = \frac{1}{\cosh(2qt)\cosh^2(qr)} \quad,$$

$$8\pi T_0^0 = 4q^2\left[\frac{3\tanh^2(2qt)+2\tanh^2(qr)-1}{\cosh^2(2qt)\cosh^8(qr)}\right] \quad, \qquad (4.28)$$

$$8\pi T_0^1 = -16q^2\left[\frac{\sinh(2qt)\sinh(qr)}{\cosh^3(2qt)\cosh^9(qr)}\right] \quad,$$

$$8\pi T_1^1 = -2q^2\left[\frac{2\tanh^2(2qt)+7\tanh^2(qr)-1}{\cosh^2(2qt)\cosh^8(qr)}\right] \quad,$$

$$8\pi T_2^2 = -2q^2\left[\frac{2\tanh^2(2qt)-\tanh^2(qr)-1}{\cosh^2(2qt)\cosh^8(qr)}\right] \quad,$$

$$8\pi T_3^3 = -4q^2\left[\frac{\tanh^2(2qt)-\tanh^2(qr)}{\cosh^2(2qt)\cosh^8(qr)}\right] \quad,$$

$$K = \frac{192q^4}{\cosh^8(2qt)\cosh^{20}(qr)}[8\cosh^2(2qt)\sinh^2(2qt)\cosh^4(qr) + 6\cosh^4(qr) - 14\cosh^4(2qt)\cosh^2(qr) + 9\cosh^4(2qt)] \quad.$$

Here q is a constant with the physical dimension of an inverse length, and the Kretschmann scalar is $192q^4$ at the origin of space and time.

The same result holds for all members of this class. Of course, the existence of unusual 4D solutions without singularities, and their 5D analogs like those described by Eqs. (4.28), does not alter the fact that much evidence indicates some kind of birth event for the universe, at least in spacetime.

4.5. Shell-like and Flat Systems

Irrespective of whether there was a big-bang singularity, it is widely believed that the presently observed structure of galaxies, clusters and superclusters is due to processes that happened in an early vacuum-dominated period of the universe. These processes could have included phase changes and the formation of bubbles and other particle-like objects, and are essentially quantum in nature. However, we need corresponding descriptions in classical relativity, especially for later epochs. Let us therefore examine shell-like systems in induced-matter theory.

Specifically, we take a metric of the form called canonical [17] and give a typical class of shell-like solutions [18]. We will study general implications of the canonical metric in Chapter 7, but for now we consider a metric of the form

$$dS^2 = \frac{\ell^2}{L^2}[A^2 dt^2 - B^2 dr^2 - C^2 r^2 d\Omega^2] - d\ell^2 . \tag{4.29}$$

Here L is a constant introduced for dimensional consistency, and the metric coefficients A, B, C depend on r only. For Eq. (4.29), the nonzero components of the 5D Ricci tensor are:

$$R_{00} = \frac{A^2}{B^2}\left\{\left[\frac{A''}{A} + \frac{A'}{A}\left(-\frac{B'}{B} + \frac{2C'}{C} + \frac{2}{r}\right) + \frac{3B'^2}{L^2}\right]\right.$$
$$\left. + \frac{\ell^2 B^2}{L^2}\left[\frac{\overset{**}{A}}{A} + \frac{1}{\ell}\left(\frac{5\overset{*}{A}}{A} + \frac{\overset{*}{B}}{B} + \frac{2\overset{*}{C}}{C}\right) + \frac{\overset{*}{A}}{A}\left(\frac{\overset{*}{B}}{B} + \frac{2\overset{*}{C}}{C}\right)\right]\right\} ,$$

$$R_{11} = -\frac{A''}{A} - \frac{2C''}{C} + \frac{B'}{B}\left(\frac{A'}{A} + \frac{2C'}{C}\right) + \frac{2}{r}\left(\frac{B'}{B} - \frac{2C'}{C}\right) - \frac{3B^2}{L^2}$$
$$- \frac{\ell^2 B^2}{L^2}\left[\frac{\overset{**}{B}}{B} + \frac{1}{\ell}\left(\frac{\overset{*}{A}}{A} + \frac{5\overset{*}{B}}{B} + \frac{2\overset{*}{C}}{C}\right) + \frac{\overset{*}{B}}{B}\left(\frac{\overset{*}{A}}{A} + \frac{2\overset{*}{C}}{C}\right)\right] ,$$

$$R_{22} = -r^2 \frac{C^2}{B^2}\left\{\left[\frac{C''}{C} + \frac{C'}{C}\left(\frac{A'}{A} - \frac{B'}{B} + \frac{C'}{C}\right)\right.\right.$$
$$+ \frac{1}{r}\left(\frac{A'}{A} - \frac{B'}{B} + \frac{4C'}{C}\right) + \frac{1}{r^2}\left(1 - \frac{B^2}{C^2}\right) + \frac{3B^2}{L^2}\right]$$
$$\left.\left.+ \frac{\ell^2 B^2}{L^2}\left[\frac{\overset{**}{C}}{C} + \frac{1}{\ell}\left(\frac{\overset{*}{A}}{A} + \frac{\overset{*}{B}}{B} + \frac{6\overset{*}{C}}{C}\right) + \frac{\overset{*}{C}}{C}\left(\frac{\overset{*}{A}}{A} + \frac{\overset{*}{B}}{B} + \frac{\overset{*}{C}}{C}\right)\right]\right\},$$

$$R_{14} = -\frac{\overset{*}{A'}}{A} - \frac{2\overset{*}{C'}}{C} + \frac{\overset{*}{B}}{B}\left(\frac{A'}{A} + \frac{2C'}{C}\right) + \frac{2}{r}\left(\frac{\overset{*}{B}}{B} - \frac{\overset{*}{C}}{C}\right), \tag{4.30}$$

$$R_{44} = -\left(\frac{\overset{**}{A}}{A} + \frac{\overset{**}{B}}{B} + \frac{2\overset{**}{C}}{C}\right) - \frac{2}{\ell}\left(\frac{\overset{*}{A}}{A} + \frac{\overset{*}{B}}{B} + \frac{2\overset{*}{C}}{C}\right).$$

Here a prime denotes $\partial/\partial r$ and an asterisk denotes $\partial/\partial \ell$ as usual. It may be verified that the components of Eq. (4.30) are zero for:

$$A = k_1\left(1 - \frac{r^2}{L^2}\right)^{1/2} + \frac{k_2 L}{\ell},$$

$$B = \frac{1}{(1 - r^2/L^2)^{1/2}}, \tag{4.31}$$

$$C = 1 + \frac{k_3 L^2}{r\ell}.$$

Here k_1, k_2 and k_3 are arbitrary dimensionless constants. The other constant L is a length in the case where the coordinates r and ℓ are lengths, but in principle it can be suppressed using an appropriate choice of units via $L = 1$. The same cannot be done for k_1, k_2 and k_3, so Eqs. (4.31) form a three-parameter class.

The properties of matter associated with Eqs. (4.31) can be evaluated using the usual method. However, for metrics of canonical form there is a more convenient formalism [17]. In terms of the 4D metric tensor $g_{\mu\nu}$ and its associated quantity $A_{\mu\nu} \equiv \overset{*}{g}_{\mu\nu}/2$, the induced energy-momentum tensor is

$$8\pi T_{\mu\nu} = \frac{3}{L^2}g_{\mu\nu} - \frac{\ell^2}{L^2}\left[\overset{*}{A}_{\mu\nu} + \left(\frac{4}{\ell} + A^\alpha_\alpha\right)\left(A_{\mu\nu} - \frac{1}{2}g_{\mu\nu}A^\beta_\beta\right)\right.$$

$$-2\left(A^{\alpha}_{\mu}A_{\nu\alpha} - \frac{1}{4}g_{\mu\nu}A^{\alpha\beta}A_{\alpha\beta}\right) . \tag{4.32}$$

For the metric (4.29), we have $A^{\alpha}_{\alpha} = \overset{*}{A}/A + 2\overset{*}{C}/C$ and $A^{\alpha\beta}A_{\alpha\beta} = \left(\overset{*}{A}/A\right)^2 + 2\left(\overset{*}{C}/C\right)^2$. It is also useful to define

$$F \equiv 1 - \frac{r^2}{L^2} ,$$

$$\begin{aligned}
H &\equiv \frac{3}{L^2} + \frac{\ell^2}{L^2}\left[\frac{\overset{*}{C}^2}{C^2} + \frac{2\overset{**}{A}\overset{*}{C}}{AC} + \frac{2}{\ell}\left(\frac{\overset{*}{A}}{A} + \frac{2\overset{*}{C}}{C}\right)\right] \\
&= \frac{1}{L^2}\left(\frac{1}{C^2} + \frac{2k_1\sqrt{F}}{AC}\right) ,
\end{aligned} \tag{4.33}$$

$$\begin{aligned}
N &\equiv \frac{\overset{*}{A}}{A} + \frac{2\overset{*}{C}}{C} + \frac{4}{\ell} \\
&= \frac{1}{\ell}\left(\frac{k_1\sqrt{F}}{A} + \frac{2}{C} + 1\right) .
\end{aligned}$$

Then Eq. (4.32) becomes

$$8\pi T_{\mu\nu} = H g_{\mu\nu} - \frac{\ell^2}{2L^2}\left[\overset{**}{g}_{\mu\nu} + N\overset{*}{g}_{\mu\nu} - g^{\alpha\beta}\overset{*}{g}_{\mu\alpha}\overset{*}{g}_{\nu\beta}\right] . \tag{4.34}$$

Evaluating this and expressing it in mixed form gives

$$8\pi T^0_0 = \frac{1}{L^2}\left(\frac{2}{C} + \frac{1}{C^2}\right) ,$$

$$8\pi T^1_1 = \frac{1}{L^2}\left(\frac{2k_1\sqrt{F}}{AC} + \frac{1}{C^2}\right) ,$$

$$8\pi T^2_2 = \frac{1}{L^2}\left(\frac{k_1\sqrt{F}}{A} + \frac{k_1\sqrt{F}}{AC} + \frac{1}{C}\right) , \tag{4.35}$$

$$T^3_3 = T^2_2 .$$

These describe a fluid which is static, spherically-symmetric and anisotropic ($T^1_1 \neq T^2_2$). But the matter is concentrated on shells: the density diverges where $C = 0$ and the pressure diverges where $A = 0$ and

$C = 0$. These latter conditions, by Eqs. (4.31), occur respectively on the surfaces defined by

$$r_A = L\left[1 - \left(\frac{k_2}{k_1}\right)^2 \left(\frac{L}{\ell}\right)^2\right]^{1/2} \, ,$$

$$r_C = \frac{|k_3|L^2}{\ell} \, .$$

(4.36)

These are the radii of surfaces where the matter properties are formally singular in a distribution which is otherwise regular (we have assumed that $\ell > 0$, $L > 0$, $k_1 k_2 < 0$ and $k_3 < 0$). The properties of the solution and its matter could be discussed at length, but we restrict ourselves to forming the 4D Ricci scalar. This is related to the trace of the energy-momentum tensor via

$$R = -8\pi T_\alpha^\alpha = -\frac{2}{L^2}\left(\frac{k_1\sqrt{F}}{A} + \frac{2k_1\sqrt{F}}{AC} + \frac{1}{C^2} + \frac{2}{C}\right) \, ,$$

(4.37)

and is obviously singular at $A = 0$, $C = 0$. The matter is concentrated on shells because of the nature of the 4D, as opposed to the 5D geometry. Indeed, the solution (4.31) is not only Ricci-flat ($R_{AB} = 0$) but may also be shown to be Riemann-flat ($R_{ABCD} = 0$). In other words, the properties of the matter are due to the fact that a curved 4D space is embedded in a flat 5D space.

Let us then look at solutions which are curved in 4D but flat in 5D, concentrating not on the cosmological case with which we are familiar but on the more general spherically-symmetric case. Such solutions are not trivial. While one can easily write down $R_{AB} = 0$ and $R_{ABCD} = 0$ in 5D, considerable skill is required in choosing a metric that gives back physically meaningful results from $G_{\alpha\beta} = 8\pi T_{\alpha\beta}$ in 4D. Presently, there is no rule known which shows how to go from a flat 5D space to a curved 4D one and have the latter possess reasonable physics. However, Abolghasem, Coley and McManus [19] used a metric in (3D) isotropic form and looked for solutions which admit the transformations (diffeomorphisms) $x^\alpha \rightarrow \bar{x}^\alpha(x^\beta)$, $x^4 \rightarrow \bar{x}^4(x^4)$, and found a class of solutions whose induced energy-momentum tensor is that of a perfect fluid and so may be applied to astrophysics. Also, Liu and Wesson [20] extended their work using a metric in standard (non-isotropic) coordinates, and found a broad class of solutions which admits a more general (anisotropic) energy-momentum tensor and may be applied to several areas of physics. We proceed to give a brief account of the latter class of solutions in order to illustrate the utility of 5D flatness.

Consider a metric which has significant dependence on the extra coordinate but whose 4D part preserves its signature under real transformations of all 5 coordinates:

$$dS^2 = A^2 dt^2 - B^2 dr^2 - C^2 d\Omega^2 - d\ell^2 \quad,$$

$$A \equiv A_1 + \ell A_2 \quad, \quad B \equiv B_1 + \ell B_2 \quad, \quad C \equiv C_1 + \ell C_2 \quad.$$

(4.38)

Here $A_1, A_2, B_1, B_2, C_1, C_2$ are functions of r. We have solved $R_{AB} = 0$ for Eq. (4.38) algebraically, but do not write out the components of this because the more numerous equations $R_{ABCD} = 0$ may be confirmed by computer. Thus a class of solutions is given by

$$A_1' = \frac{-aC_2 C_1'}{\left(1-C_2^2\right)^{1/2}} \quad, \quad A_2 = a(1 - C_2^2)^{1/2} \quad,$$

$$B_1 = \frac{C_1'}{\left(1-C_2^2\right)^{1/2}} \quad, \quad B_2 = \frac{C_2'}{\left(1-C_2^2\right)^{1/2}} \quad.$$

(4.39)

Here a is an arbitrary constant, a prime denotes $\partial/\partial r$ and $C_1(r), C_2(r)$ remain arbitrary. Of these latter two functions, one can in principle be absorbed by an appropriate choice of the r coordinate. However, while this is mathematically possible, it is often more instructive to choose C_1 and C_2 so as to give the metric and its corresponding induced matter relevance to some physical situation.

For example, if we choose $C_2 = r/L$ where L is some length and put $a = 1/L$, we obtain

$$dS^2 = \frac{\ell^2}{L^2} \left\{ \left[\left(1 - \frac{r^2}{L^2}\right)^{1/2} + \frac{L}{\ell} A_1(r) \right]^2 dt^2 - \frac{\left[1 + LC_1'/\ell\right]^2}{\left[1 - r^2/L^2\right]} dr^2 \right.$$

$$-r^2 \left[1 + \frac{L}{\ell}\frac{C_1}{r}\right]^2 d\Omega^2 \right\} - d\ell^2 \quad, \tag{4.40}$$

$$A_1 = A_0 - \frac{1}{L^2} \int \frac{r C_1' dr}{(1 - r^2/L^2)^{1/2}} \quad.$$

Here A_0 is an arbitrary constant and $C_1(r)$ is still an arbitrary function. We recognize this as a 5D generalization of the de Sitter-type solution

$$dS^2 = \frac{\ell^2}{L^2}\left[\left(1 - \frac{r^2}{L^2}\right) dt^2 - \frac{dr^2}{(1 - r^2/L^2)} - r^2 d\Omega^2\right] - d\ell^2 \quad, \tag{4.41}$$

which we recover if $C_1 = 0$ and $A_0 = 0$. (We will see in Chapter 7 that 5D canonical space embeds all 4D Einstein spaces with a cosmological constant, the latter being $3/L^2$.) If alternatively $C_1 = 0$ but $A_0 \neq 0$ then

$$dS^2 = \frac{\ell^2}{L^2} \left\{ \left[\left(1 - \frac{r^2}{L^2}\right)^{\frac{1}{2}} + \frac{A_0 L}{\ell}\right]^2 dt^2 - \frac{dr^2}{(1 - r^2/L^2)} - \right.$$

$$\left. -r^2 d\Omega^2 \right\} d\ell^2. \tag{4.42}$$

And if $C_1 = C_0 = $ constant, then Eq. (4.40) gives

$$dS^2 = \frac{\ell^2}{L^2} \left\{ \left[\left(1 - \frac{r^2}{L^2}\right)^{\frac{1}{2}} + \frac{A_0 L}{\ell}\right]^2 dt^2 - \frac{dr^2}{(1 - r^2/L^2)} \right.$$

$$\left. - r^2 \left(1 + \frac{C_0 L}{\ell r}\right)^2 d\Omega^2 \right\} - d\ell^2 \quad. \tag{4.43}$$

Inspection of Eqs. (4.42) and (4.43) shows that the physics they imply can be quite different from that of the conventional solution (4.41). Insofar as the 4D de Sitter solution contained in Eq. (4.41) is the spacetime often regarded in particle physics as defining the vacuum, it can be argued that the 5D solution (4.40) represents a 'generalized' vacuum.

4.6. Waves in de Sitter Vacuum

We saw in the preceding section that the local version of the 4D de Sitter solution (static with spherical symmetry) can be embedded in 5D canonical space and generalized. It is natural to ask if a similar procedure can be followed for the cosmological version of the 4D de Sitter solution (non-static and uniform, as discussed in Chapter 1). This can indeed be done, as shown by Billyard and Wesson [21]. If the elements of the metric tensor are allowed to take complex values, then there is a class of exact solutions to $R_{AB} = 0$ which can best be described as 4D vacuum waves. There are issues of interpretation with such solutions to which we will return later, but for now we wish to present a wave-like solution and give a physical interpretation of it.

Consider then the x^4-dependent 5D metric specified by

$$dS^2 = \ell^2 dt^2 - \ell^2 e^{i(\omega t + k_x x)} dx^2 - \ell^2 e^{i(\omega t + k_y y)} dy^2$$
$$-\ell^2 e^{i(\omega t + k_z z)} dz^2 + L^2 d\ell^2 \quad . \tag{4.44}$$

Here ω is a frequency, k_x etc. are wavenumbers, and L measures the size of the extra dimension. The nonzero components of the Riemann-Christoffel tensor for Eq. (4.44) have the form

$$R_{txtx} = -\frac{\ell^2}{4L^2}(\omega^2 L^2 - 4)e^{i(\omega t + k_x x)} \quad \text{etc. ,}$$

$$R_{xyxy} = \frac{\ell^2}{4L^2}(\omega^2 L^2 - 4)e^{i(2\omega t + k_x x + k_y y)} \quad \text{etc. ,} \tag{4.45}$$

with two similar relations in each case obtained by permutations of the space coordinates. The nonzero components of the Ricci tensor look like

$$R_{tt} = -\frac{3}{4L^2}(\omega^2 L^2 - 4) \quad ,$$

$$R_{xx} = \frac{3}{4L^2}(\omega^2 L^2 - 4)e^{i(\omega t + k_x x)} \quad \text{etc.} \tag{4.46}$$

It is clear from these relations that the field equations $R_{AB} = 0$ are satisfied for $\omega = \pm 2/L$. It is also clear from Eqs. (4.45) that the solution is 5D flat.

However, it is not flat in 4D, and this implies non-trivial physics in the STM picture. Thus the 4D Ricci or curvature scalar is

$$R = \frac{1}{4L^2} \left[\overset{*}{g}{}^{\mu\nu} \overset{*}{g}_{\mu\nu} + \left(g^{\mu\nu} \overset{*}{g}_{\mu\nu} \right)^2 \right] = \frac{12}{\ell^2 L^2} , \tag{4.47}$$

and the 4D effective energy-momentum tensor has

$$8\pi T_{00} = -\frac{3\omega^2}{4} ,$$
$$\tag{4.48}$$
$$8\pi T_{11} = \frac{3\omega^2}{4} e^{i(\omega t + k_x x)} \quad \text{etc.}$$

It is possible, as we have done in other cases, to match these components to those of a perfect fluid with $T_{\alpha\beta} = (\rho + p)u_\alpha u_\beta - p g_{\alpha\beta}$. It transpires that for the metric (4.44), the pressure and density are given by

$$8\pi p = -8\pi \rho = \frac{3\omega^2}{4\ell^2} . \tag{4.49}$$

That is, the equation of state is that of the vacuum in general relativity. In fact, the induced medium is a de Sitter vacuum which, if interpreted in terms of the cosmological constant, would have $\Lambda = -3\omega^2/4\ell^2$.

This result, while curious, turns out to be a special case of a general theorem on canonical space that will be treated fully in Chapter 7. Our metric (4.44) does in fact have what is known as canonical form, though the latter is usually written in an alternative way that is easier to interpret because it makes the cosmological constant a true constant as measured by an observer confined to the ℓ-hypersurface where Einstein's equations hold. Thus:

$$dS^2 = \frac{\ell^2}{L^2} ds^2 + \varepsilon d\ell^2 , $$
$$\tag{4.50}$$
$$\Lambda = -\frac{3\varepsilon}{L^2} .$$

Here ds^2 is the metric of *any* 4D solution of the vacuum Einstein equations. This situation actually follows from Campbell's embedding theorem (see Chapter 2). We see that $\Lambda > 0$ for a space like extra dimension ($\varepsilon = -1$) and $\Lambda < 0$ for a timelike extra dimension ($\varepsilon = +1$). The implications of Eqs. (4.50) are not confined to vacuum waves, and we will meet them in numerous places.

4.7. Conclusion

The solutions discussed in this chapter are examples, and many more remain to be found by the motivated reader. The 5D field equations for 3D spherical symmetry have numerous applications beyond the solutions noted in Section 4.2, and some will surely have interesting physics associated with them like the one of Section 4.3. That solution has good thermodynamical properties and is self-similar or scale-free like many astrophysical systems; though as we saw, both attributes require a careful consideration of coordinates to become apparent. Not all systems in the real universe are spherical, however. Most galaxies, for example, are axially symmetric, with geometries which in some cases can be approximated by the equations of Section 4.4. Many clusters of galaxies are now known to be more flattened than elliptical galaxies, while structure on a still larger scale consists of filaments and voids with a variety of shapes that structure is widely believed to have grown over time by gravitational instability, around seed perturbations left over from the very early universe. The original perturbations may have been connected with the growth of bubbles in a vacuum-dominated phase. In any event, there is a need for models with shell-like structure, like those considered in Section 4.5. Many such solutions turn out to be 5D-flat, and the field equations can be tackled from this as a starting point.

Another motivation for studying 5D-flat solutions is to match the standard class of 5D cosmologies, which we found was flat in Section 3.2. Assuming that the early universe was indeed dominated by vacuum energy, it is logical to take the 4D metric to be that of de Sitter. This, as we saw in Chapter 1, comes in two forms, with a positive and negative cosmological constant. However, as we will find out in Chapter 7, the extension of the de Sitter space to 5D gives a canonical space where the so-called cosmological 'constant' is actually a variable measure of the energy density of the vacuum. It is even possible that small, stable regions of ordinary 3D space with negative vacuum energy are embedded in an expanding background with positive vacuum energy. This may be a relatively crude model for particles in the universe; but it is certain that modern quantum theory makes the vacuum into a much richer place than the sterile emptiness of classical physics. One remarkable result of the extension of the classical de Sitter solution to 5D is the existence of vacuum waves, as outlined in Section 4.6. We will come back to these waves in various places, including Chapter 8, which deals with the application of 5D to particles.

The present chapter has been mostly concerned with the application of the 5D field equations to finding solutions whose properties of matter can be derived by the standard Space-Time-Matter method discussed in Chapter 2. That method is of comparatively recent vintage (1992), and the complimentary approach to 5D known as Membrane Theory dates from the same decade (1998, 1999). However, there is a class of 5D objects whose history goes back significantly further (1970). These are the so-called solitons, about which so much has been written that they require a chapter to themselves.

References

[1] Billyard, A. and Wesson, P.S., *Phys. Rev.* **D53**, 731 (1996a).
[2] Henriksen, R.N. and Wesson, P.S., *Astrophys. Space Sci.* **53**, 429 (1978).
[3] Ponce de Leon, J. and Wesson, P.S., *Int. J. Mod. Phys.* **D15**, 395 (2006).
[4] Gross, D.J. and Perry, M.J., *Nucl. Phys.* **B226**, 29 (1983).
[5] Myers, R.C. and Perry, M.J., *Ann. Phys.* (New York) **172**, 304 (1986).
[6] Clement, G., *Gen. Rel. Grav.* **18**, 137 (1986).
[7] Bruckman, W., *Phys. Rev.* **D34**, 2990 (1986); **D36**, 3674 (1987).
[8] Frolov, V.P., Zelnikov, A.I. and Bleyer, U., *Ann. Phys.* (Leipzig) **44**, 371 (1987).
[9] Horne, J.H. and Horowitz, G.T., *Phys. Rev.* **D46**, 1340 (1992).
[10] Matos, T., *J. Math. Phys.* **35**, 1302 (1994).
[11] Senovilla, J.M.M., *Phys. Rev. Lett.* **64**, 2219 (1990).
[12] Chinea, F.J., Fernandez-Jambrina, L. and Senovilla, J.M.M., *Phys. Rev.* **D45**, 481 (1992).
[13] Ruiz, E. and Senovilla, J.M.M., *Phys. Rev.* **D45**, 1995 (1992).
[14] Tikekar, R., Patel, L.K. and Dadhich, N., *Gen. Rel. Grav.* **26**, 647 (1994).

[15] Banerjee, A., Das, A. and Panigrahi, D., *Phys. Rev.* **D51**, 6816 (1995).

[16] Chatterjee, S., Wesson, P.S. and Billyard, A., *Nuovo Cim.* **B112**, 1667 (1997).

[17] Mashhoon, B., Liu, H. and Wesson, P.S., *Phys. Lett. B* **331**, 305 (1994).

[18] Wesson, P.S. and Liu, H., *Phys. Lett. B* **432**, 266 (1998).

[19] Abolghasem, G., Coley, A.A. and McManus, D.J., *J. Math. Phys.* **37**, 361 (1996).

[20] Liu, H. and Wesson, P.S., *Gen. Rel. Grav.,* **30**, 509 (1998).

[21] Billyard, A. and Wesson, P.S., *Gen. Rel. Grav.* **28**, 129 (1996b).

Chapter 5

The Solitons

5.1. Introduction

Solitons are solutions of the 5D Ricci-flat field equations $R_{AB} = 0$, where the metric is static, spherically symmetric in ordinary space and asymptotically flat [1]. These properties are shared by the unique Schwarzschild solution of four-dimensional general relativity. However, it is a mistake to confuse the two. We wish to clear up this and other misunderstandings about the physical nature of solitons, and will conclude that in general they possess gravitational and scalar 'charges' which can be regarded as bivalent aspects of mass.

Section 2 gives a short summary of the long and convoluted history of the solitons, to establish what is already known about them. Section 3 focuses on their physical properties, and offers some insights about their real nature. Section 4 derives a class of time-dependent solitons. Section 5 offers some suggestions about how solitons might in principle be detected. Section 6 is as usual a conclusion.

5.2. Historical Survey

What is now known as the 5D soliton metric was first obtained by Kramer in 1970 as an extension of the 4D Schwarzschild solution [2]. It was later derived independently by Sorkin [3], who was convinced it described a magnetic monopole, an interpretation adopted by Gross and Perry [4], who however also realized that it involved two kinds of mass. Then Davidson and Owen derived it under the belief it described a kind of black hole [5], a view criticized by Dereli [6]. Related metrics were studied by Sokolowski and Carr [7] and Liu [8], and a version of the soliton was used to model a wormhole by Agnese *et al.* [9]. The soliton metric or special cases of it were subsequently discovered again by other workers [10, 11]; but there was no consensus about the physical meaning of the metric.

This situation changed in 1992, when Wesson calculated an effective source for the soliton metric [12]. His method was also applied to cosmology [13], and developed with Ponce de Leon into a general scheme to derive the 4D physical source for any 5D metric ([14, 15]; see Chapter 2). Together, these workers applied the Tolman-Whittaker integral to calculate the effective 4D gravitational mass of a soliton [16]. This is the source which determines the orbit of a test particle in the field of the soliton.

By contrast, what is commonly termed the 'energy' of the soliton may be calculated using only the metric, without regard to any matter that may be associated with it. This metric-based energy was considered by Gross and Perry [4] and discussed at length by Bombelli *et al.* [17]; but a later calculation by Deser and Soldate [18] gave a slightly discrepant answer. The subject was taken up again by Billyard *et al.* [19], who confirmed the earlier result. It was further confirmed by Sajko and Wesson [20, 21], who used a Hamiltonian method to calculate the metric-based energy of electrically-charged solitons.

The extension of the soliton class of metrics to include electric charge had been made slightly earlier ([22, 23]; see Chapter 6). However, it should be mentioned that a different approach to charged 5D objects had been made in 1985 by Nodvik [24], who in a little-noticed paper argued that such objects could cause 4D vacuum polarization. This subject remains controversial, but should be examined in detail because Liko has shown that 5D solutions can have measurable electromagnetic effects [25], and it remains an open question whether the fine-structure constant varies with cosmological time.

There are other controversial aspects of 5D relativity, especially in regard to the basic (uncharged) solitons. Ponce de Leon has made intensive efforts to resolve some of these. He has related the solitons to an extra symmetry of the (3D) spherically-symmetric field equations [26], looked at the status of the Weak Equivalence Principle in 5D [27], and made a detailed investigation of the two parameters which define the soliton class of solutions [28]. The last study agrees for the most part with an investigation of the geometrical properties of solitons by Lake [29]. These recent investigations are mainly concerned with theoretical properties of the 5D soliton class of solutions, and in differentiating it from the 4D Schwarzschild solution.

Experimental information about solitons should also be mentioned, even though it dates from the time when they were thought to be matter-free extensions of the unique 4D solution. A comprehensive comparison with the classical solar-system tests was made by Kalligas *et al.* [30]. This showed that the two soliton parameters must be very close to the values which give back the Schwarzschild solution. The two parameters concerned obey a quadratic consistency relation which will be examined in detail in Section 3. All researchers agree that this consistency relation is crucial to an understanding of the solitons, but opinions differ as to how it should be most usefully presented. In an alternative investigation of the solar-system tests, Lim *et al.* [31] suggested that the relation could be cast in the form of the equation for an ellipse. This possibility was made explicit by Ponce de Leon [28], and a different version of this will be presented in Section 5.3. In regard to experimental work on the solitons, it was suggested that the classical tests could be augmented by investigating the precession of a spinning object in Earth orbit, as in the Gravity Probe-B gyroscope experiment. The basic relations for this were worked out by Liu and Wesson [32], but the magnitude of the difference from general relativity was too small to be detected by that experiment. More general studies of this effect and the other solar-system tests were made by Liu and Overduin [33, 34]. They showed that in a comparison of the 5D soliton metric with the 4D Schwarzschild metric, the bodies of the solar system agree with the latter to high accuracy, in some cases better than 1 part in 10^6.

This remarkable level of support for the 4D theory over the 5D seems to call for explanation. One such explanation was provided in 1994 by Mashhoon *et al.* [35], who found a new 5D embedding for the 4D Schwarzschild solution. The so-called canonical embedding actually leads to dynamical results in 5D and 4D which are *identical*. This can be traced to the fact that the canonical metric, unlike the soliton one, has a flat extra dimension and a quadratic prefactor in the extra coordinate which multiplies onto 4D spacetime. Indeed, it is a theorem that the 5D canonical metric smoothly embeds *any* 4D vacuum metric of general relativity. This in turn follows from Campbell's theorem, which ensures that any solution of the 5D Ricci-flat equations in apparent vacuum embeds a corresponding solution of the 4D Einstein equations, usually with matter.

The canonical metric will be discussed in Chapter 7, and Campbell's Theorem was outlined in Chapter 2. For present purposes, it is important to recall that Campbell's theorem lay forgotten in the literature from the 1920s to the 1990s, when Tavakol and coworkers resurrected it [36-38]. Further work followed [39], and it became clear that the solitons are not vacuum solutions, but characterized by the matter distributions originally calculated by Wesson [12]. The problem of 'finely tuned' parameter values in the case of the soliton metric disappears when the canonical metric is used instead. The inference is that the true embedding of the 4D Schwarzschild solution is not provided by the soliton metric but by the corresponding canonical metric.

5.3. Properties of Static Solitons

Given that the solitons are really not versions of the Schwarzschild solution, then what are they? We proceed to answer this by using the STM formalism to elucidate their properties.

The metric for the solitons in isotropic coordinates is given by

$$dS^2 = \left(\frac{1-M/2r}{1+M/2r}\right)^{2a} dt^2 - \left(\frac{1-M/2r}{1+M/2r}\right)^{2(1-a-b)} \left(1+\frac{M}{2r}\right)^4 [dr^2 +$$
$$r^2 d\Omega^2] - \left(\frac{1-M/2r}{1+M/2r}\right)^{2b} d\ell^2 \quad . \tag{5.1}$$

Here $d\Omega^2 \equiv (d\theta^2 + \sin^2\theta d\phi^2)$ as usual, and the extra dimension is taken to be spacelike. There are two dimensionless parameters in the above metric, which satisfy a consistency relation set by the spatial part of the 5D field equations $(R_{ij} = 0, i \neq j)$. This reads

$$a^2 + ab + b^2 = 1 \quad . \tag{5.2}$$

This relation is invariant under exchange of a and b, and a change in the sign of both. These properties also follow from the field equations, and the extra symmetry in them discussed by Ponce de Leon [26]. The terminology for the two parameters varies between workers, as does their assumed form for the metric, and to assist readers in translating between different usages a note is added in the bibliography at the end of this chapter [40]. In the metric (5.1), the source is denoted for reasons of convention by M; but as will become clear below it should *not* be identified with a simple mass at the centre of the geometry, as in the isotropic form of the 4D Schwarzschild solution. The latter, plus an extra flat piece, is recovered from Eq. (5.1) when $a = 1, b = 0$. Also, the centre of the spatial geometry is now at $r = M/2$, and the temporal part of the geometry goes to zero at the same place. This means that the solitons in general do not have horizons of the conventional sort, as remarked by several workers. That is, solitons should not be called black holes.

Furthermore, the solitons have matter associated with them. To see this, recall that they are solutions of the 5D field equations $R_{AB} = 0$, and if they are also to be solutions of the 4D Einstein equations $G_{\alpha\beta} = 8\pi T_{\alpha\beta}$ then the former set of equations has to be systematically reduced to the latter set. The procedure for doing this was specified in Chapter 2, and applied to Eq. (5.1) it gives the effective or induced energy-momentum tensor $T_{\alpha\beta}$ for solitons. From this, we find that the density and the components of the (anisotropic) pressure are given by

$$8\pi\rho = \frac{-abM^2r^4}{(r-M/2)^4(r+M/2)^4} \left(\frac{r-M/2}{r+M/2}\right)^{2(a+b)} , \tag{5.3a}$$

$$8\pi p_r = \frac{2bMr^3}{(r-M/2)^3(r+M/2)^3} \left(\frac{r-M/2}{r+M/2}\right)^{2(a+b)}$$

$$+ \frac{bM^2r^4(a+2b-4r/M)}{(r-M/2)^4(r+M/2)^4}\left(\frac{r-M/2}{r+M/2}\right)^{2(a+b)} , \qquad (5.3b)$$

$$8\pi p_{\theta,\phi} = \frac{-bMr^3}{(r-M/2)^3(r+M/2)^3}\left(\frac{r-M/2}{r+M/2}\right)^{2(a+b)}$$
$$- \frac{bM^2r^4(a+b-2r/M)}{(1-M/r)^4(1+M/r)^4}\left(\frac{r-M/2}{r+M/2}\right)^{2(a+b)} . \qquad (5.3c)$$

These components satisfy the equation of state for radiation or ultra-relativistic particles. Defining $\bar{p} \equiv (p_r + p_\theta + p_\phi)/3$, this is $\bar{p} = \rho/3$. It is in fact a theorem that for 5D metrics independent of x^4 the components of T_β^α satisfy

$$T \equiv T_\alpha^\alpha = T_0^0 + T_1^1 + T_2^2 + T_3^3 = \rho - p_r - p_\theta - p_\phi = 0 . \quad (5.4)$$

For $T_0^0 = \rho > 0$, Eq. (5.3a) above shows that the dimensionless parameters a and b must have opposite signs.

The gravitational mass of the fluid defined by Eqs. (5.3) is given by the Tolman-Whittaker formula, which is an appropriately-defined integral over $(T_0^0 - T_1^1 - T_2^2 - T_3^3) = (\rho + p_r + p_\theta + p_\phi)$. It is

$$M_g(r) = aM \left(\frac{r-M/2}{r+M/2}\right)^{-b} . \qquad (5.5)$$

For this to be positive requires $a > 0$, so by the preceding condition for positive density it is necessary that $b < 0$. Then the exponent in Eq. (5.5) is positive, and $M_g(r) \to 0$ for $r \to M/2$. This is the centre of the geometry as defined above, so the gravitational mass is zero at the physical centre of ordinary space. Also $M_g(r) \to aM$ for $r \to \infty$, so the asymptotic 5D mass is in general not equal to the standard 4D mass. The Schwarzschild case ($a = 1, b = 0$) is exceptional, and from the viewpoint of functional analysis not strictly a member of the soliton class at all. In general, solitons are spherically-symmetric clouds of radiation-like matter whose properties fall off rapidly with distance ($\propto 1/r^4$) but extend indefinitely far into space.

The gravitational mass $M_g(r)$ in Eq. (5.5) is obtained from an integral over what in cosmology is called the gravitational density $(\rho + 3p)$. This is the appropriate density for calculating the attraction of a portion of a continuous fluid in accordance with the radial equation of motion (also known in this application as the Raychaudhuri equation). There is also the so-called inertial density $(\rho + p)$, which governs the mechanical properties of matter in accordance with the equation of continuity. For radiation-like matter, the equation of state is given by $T = 0$ or $\bar{p} = \rho/3$. Then the gravitational and inertial densities are 2ρ and $4\rho/3$ respectively. This means that for the solitons, their effective inertial mass is $M_i(r) = (2/3)M_g(r)$. Insofar as the Weak Equivalence Principle is sometimes loosely stated as the proportionality of gravitational and inertial mass, solitons obey it. Of course, most ordinary matter in the universe (as opposed to radiation or vacuum) has $p \ll \rho$, so the gravitational and inertial masses are equal to a good approximation. A more sophisticated discussion of 5D objects and the Weak Equivalence Principle has been given by Ponce de Leon [27]; and the astrophysical implications of solitons will be discussed in Section 5.5 below.

The physical nature of solitons becomes clearer under a change from isotropic to quasi-Schwarzschild coordinates via $r \to \left(r - M + \sqrt{r(r-2M)}\right)/2$. Then Eq. (5.1) becomes

$$dS^2 = \left(1 - \frac{2M}{r}\right)^a dt^2 - \frac{dr^2}{\left(1-\frac{2M}{r}\right)^{a+b}} - \frac{r^2 d\Omega^2}{\left(1-\frac{2M}{r}\right)^{a+b-1}}$$
$$- \left(1 - \frac{2M}{r}\right)^b d\ell^2 \ . \tag{5.6}$$

From this, it is obvious that for $(2M/r) \ll 1$ what would normally be called *the* mass is split into parts aM and bM, which figure in the first and last terms and are frequently called the gravitational mass and the inertial mass. The former appellation is justified in that $M_g(r) \to aM$ for $r \gg M/2$ in Eq. (5.6). But the inertial mass of a source means different things to different people, so it is better to refer to bM as the scalar mass. Irrespective of what we call the components of the source, it should be recalled that the actual matter associated with it is the fluid whose density and pressure are given by Eq. (5.3). That fluid has the equation of state $\bar{p} = \rho/3$; but it should not be automatically assumed that it consists of photons or an electromagnetic field, since the noted equation of state holds for any particles that are massless and move at the speed of light. Indeed, they could be gravitons and scalarons, since the metric (5.6) describes gravitational and scalar fields.

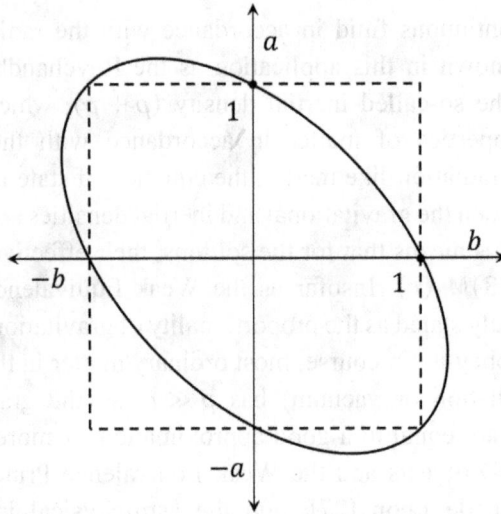

Figure 5.1: The consistency relation $a^2 + ab + b^2$ for 5D solitons, with a square of half-side unity for reference (dashed). The noted relation describes a tilted ellipse, with semi-major axis $\sqrt{2}$ and semi-minor axis $\sqrt{2/3}$.

More insight into solitons can be obtained by plotting the consistency relation (5.2) for their defining parameters, as in Figure 5.1. Allowed values of a and b must lie on the tilted ellipse shown there. However, physical constraints ($a > 0, b < 0$) mean that attention should be focused on the upper left quadrant of the plot. The ellipse defined by Eq. (5.2) is tilted because of the cross-term in that relation.

The question arises of whether the ellipse can be rotated into its standard configuration. This can be accomplished by a coordinate transformation on the metric according to Dereli [6], or more directly by a redefinition of the parameters. It may be verified that the appropriate relations are:

$$\alpha = k_1 a + k_2 b \quad , \quad \beta = k_2 a + k_1 b \quad ,$$

$$a = \frac{\alpha k_1 - \beta k_2}{(k_1^2 - k_2^2)} \quad , \quad b = \frac{\beta k_1 - \alpha k_2}{(k_1^2 - k_2^2)} \quad , \tag{5.7}$$

$$k_1 = \pm \left(\frac{1}{2} \pm \frac{\sqrt{3}}{4} \right)^{1/2} \quad , \quad k_2 = \frac{1}{4} \left(\frac{1}{2} \pm \frac{\sqrt{3}}{4} \right)^{-1/2} \quad .$$

With the new parameters α, β in place of the old ones a, b the consistency relation (5.2) reads

$$\alpha^2 + \beta^2 = 1 \ . \tag{5.8}$$

This is the equation of an ellipse in standard rectangular coordinates x, y where $\alpha \equiv x/A, \beta \equiv y/B$ and A, B are the semi-axes. Alternatively, the ellipse may be considered as the locus of the tip of a moving, right-angled triangle; and then Eq. (5.8) is just the statement that the sides (squared) sum to the hypotenuse (squared) with value unity. Alternatively, the parameters above may be replaced by writing $\alpha = \sin\theta$ and $\beta = \cos\theta$, so that Eq. (5.8) becomes the familiar relation $\sin^2\theta + \cos^2\theta = 1$. This way of describing the properties of the solitons is the most straightforward. We see that, in a sense, the gravitational and scalar sources are parts of the same thing. Or, there is only one source, which is bivalent.

It is tempting, having arrived at this conclusion, to try and visualize the 'thing' at the centre of the geometry. This, however, probably makes no sense. We recall that while a soliton may be regarded as a spherically-symmetric cloud of massless particles with the density and pressure given by Eqs. (5.3), its gravitational mass (5.5) goes to zero at the centre. This centre is at $r = M/2$ in isotropic coordinates with the metric (5.11), and at $r = 2M$ in quasi-Schwarzschild coordinates with the metric (5.6). We have calculated the 5D Kretschmann scalar $K \equiv R_{ABCD}R^{ABCD}$ and associated 4D curvature scalar $C \equiv R_{\alpha\beta}R^{\alpha\beta}$ for both sets of coordinates [15,16]. Both diverge at the centre. Thus, the gravitational and scalar fields of a soliton emerge from a naked singularity.

5.4. Time-Dependent Solitons

The solitons studied in the preceding section are independent of $x^0 = t$ and $x^4 = \ell$, but in the absence of a Birkhoff-type theorem they can be generalized in both regards [41, 42]. The time-dependent extension is particularly interesting, because if static solitons exist in nature they presumably formed by some dynamical process. Also, time-dependent solutions which represent centrally-condensed clouds of hot matter may be applicable to the problem of galaxy formation. Thus we will give a particularly interesting class of time-dependent soliton solutions found by Wesson *et al.* [43] and evaluate their properties of matter.

Let us consider the time-dependent generalization of the static soliton metric (5.1) in the form

$$dS^2 = A^2(t,r)dt^2 - B^2(t,r)dx^i dx^i - C^2(t,r)d\ell^2. \quad (5.9)$$

We are here for algebraic convenience using Cartesian coordinates for the 3-space, which are related to spherical polar coordinates by the usual relations $(x^1 = r\sin\theta\cos\phi, \ x^2 = r\sin\theta\sin\phi, \ x^3 = r\cos\theta)$. For the metric (5.9), the nonzero components of the 5D Ricci tensor are

$$R_{00} = \frac{AA'}{B^2}\left(\frac{2}{r} + \frac{A''}{A'} + \frac{B'}{B} + \frac{C'}{C}\right) + \frac{\dot{A}}{A}\left(\frac{3\dot{B}}{B} + \frac{\dot{C}}{C}\right) - \left(\frac{3\ddot{B}}{B} + \frac{\ddot{C}}{C}\right) \ ,$$

$$R_{0i} = \left(\frac{x^i}{r}\right)\left[\frac{\dot{B}}{B}\left(\frac{2A'}{A} + \frac{2B'}{B} + \frac{C'}{C}\right) + \frac{A'\dot{C}}{AC} - \frac{2B'}{B} - \frac{\dot{C'}}{C}\right] \ ,$$

$$\begin{aligned}
R_{ij} = -\delta_{ij}&\left[-\frac{B\dot{B}}{A^2}\left(\frac{\ddot{B}}{\dot{B}} + \frac{2\dot{B}}{B} - \frac{\dot{A}}{A} + \frac{\dot{C}}{C}\right)\right.\\
&\left.+\frac{1}{r}\left(\frac{A'}{A} + \frac{3B'}{B} + \frac{C'}{C}\right) + \frac{B''}{B} + \frac{B'}{B}\left(\frac{A'}{A} + \frac{C'}{C}\right)\right]\\
&-\frac{x^i x^j}{r^2}\left[\frac{A''}{A} + \frac{B''}{B} + \frac{C''}{C} - \left(\frac{1}{r} + \frac{2B'}{B}\right)\left(\frac{A'}{A} + \frac{B'}{B} + \frac{C'}{C}\right)\right] \ ,
\end{aligned}$$

$$R_{44} = -\frac{CC'}{B^2}\left(\frac{2}{r} + \frac{A'}{A} + \frac{B'}{B} + \frac{C''}{C'}\right) + \frac{C\dot{C}}{A^2}\left(-\frac{\dot{A}}{A} + \frac{3\dot{B}}{B} + \frac{\ddot{C}}{C}\right) \ .$$

(5.10)

Using these, we look for solutions of $R_{AB} = 0$ that are separable:

$$A = \alpha(t)a(r) \ , \quad B = \beta(t)b(r) \ , \quad C = \gamma(t)c(r) \ . \quad (5.11)$$

There are now six functions to be determined, but from Eq. (5.9) we see that a coordinate transformation on t allows us to set $\alpha(t) = 1$ without loss of generality. Also, let us suppose that $a(r), b(r), c(r)$ are the same functions as in the static case. Then it may be verified that Eqs. (5.10) and (5.11) give four relations. Together with the consistency relation (which is the same as before), these have a unique solution. It reads

$$dS^2 = \left(\frac{ar-1}{ar+1}\right)^{4/\sqrt{3}} dt^2 - \left(\frac{a^2r^2-1}{a^2r^2}\right)^2 \left(\frac{ar+1}{ar-1}\right)^{2/\sqrt{3}} (1+Ht)dx^i dx^i$$
$$- \left(\frac{ar+1}{ar-1}\right)^{2/\sqrt{3}} (1+Ht)^{-1}d\ell^2 \quad . \tag{5.12}$$

Here H is a constant with the physical dimensions of inverse time, like Hubble's parameter of cosmology. And a is a constant we have used above to specify the strength of the source, in place of the mass M (it should not be confused with previous usages). However, $a \equiv 2/M$, and we see that for $a \to \infty$ ($M \to 0$) the metric (5.12) gives back the Robertson-Walker one for radiation.

The matter associated with Eq. (5.12), for general values of the constant a, can be obtained by using the 5D $\to$ 4D reduction mechanism specified in Chapter 2. The resulting components of the induced energy-momentum tensor are, in mixed form:

$$8\pi T_0^0 = \frac{1}{(1+Ht)} \frac{8a^6r^4}{3(a^2r^2-1)^4} \left(\frac{ar-1}{ar+1}\right)^{\frac{2}{\sqrt{3}}} + \frac{3H^2}{4(1+Ht)^2} \left(\frac{ar-1}{ar+1}\right)^{-\frac{4}{\sqrt{3}}} \quad ,$$

$$8\pi T_0^1 = -\frac{H}{(1+Ht)^2} \frac{4a}{\sqrt{3}} \frac{a^4r^4}{(a^2r^2-1)^3} \left(\frac{ar-1}{ar+1}\right)^{\frac{2}{\sqrt{3}}} \quad ,$$

$$8\pi T_1^1 = -\frac{1}{(1+Ht)} \frac{4a^5r^3}{\sqrt{3}(a^2r^2-1)^4} \left(\frac{ar-1}{ar+1}\right)^{\frac{2}{\sqrt{3}}} (a^2r^2 + 1)$$
$$- \frac{H^2}{4(1+Ht)^2} \left(\frac{ar-1}{ar+1}\right)^{-\frac{4}{\sqrt{3}}} \quad , \tag{5.13}$$

$$8\pi T_2^2 = -\frac{1}{(1+Ht)} \frac{2a^5r^3}{\sqrt{3}(a^2r^2-1)^4} \left(\frac{ar-1}{ar+1}\right)^{\frac{2}{\sqrt{3}}} \left(\frac{2}{\sqrt{3}} ar - a^2r^2 - 1\right)$$

$$- \frac{H^2}{4(1+Ht)^2} \left(\frac{ar-1}{ar+1}\right)^{-\frac{4}{\sqrt{3}}} \quad ,$$

$$T_3^3 = T_2^2 \quad .$$

These components may be shown to obey $T_\alpha^\alpha = 0$, so the equation of state of the fluid is radiation-like as expected. However, the pressure is anisotropic, though its average value is related to the density by

$$\rho = 3\bar{p} = \frac{1}{(1+Ht)}\frac{8a^6r^4}{3(a^2r^2-1)^4}\left(\frac{ar-1}{ar+1}\right)^{\frac{2}{\sqrt{3}}} + \frac{3H^2}{4(1+Ht)^2}\left(\frac{ar+1}{ar-1}\right)^{\frac{4}{\sqrt{3}}} . \quad (5.14)$$

The first term of this expression is similar to the static case, but multiplied by a time factor. The second term represents a contribution to the density and pressure from the motion. Thus we have a natural generalization of the properties of matter of the static soliton.

5.5. Detecting Solitons

While 5D relativity has a limit in which it reduces smoothly to the 4D one, solitons involve the fifth dimension in a meaningful manner, and thereby provide a possible way of deciding between 4D and 5D. But while STM theory predicts solitons, it does not provide detailed information on things like their temperature and spectrum. We do not even know how 'big' a soliton might be.

Astrophysical solitons, however, find a natural niche in that they could in principle provide some of the dark matter or dark energy inferred from the observations of galaxies [44, 45]. There are three ways that might be employed to detect them.

(a) Background radiation, especially in the microwave part of the spectrum, would be perturbed by the extended radiation-like profiles of solitons. Consider a population of solitons spread approximately uniformly through ordinary space. Because of their nature, the nearest one to an observer tends to dominate. That soliton has a gravitational mass $M_g(r)$ as a function of distance from its centre given by Eq. (5.5) and the density of its radiation-like matter varies as M_g^2/r^4 by Eq. (5.3a). Suppose that this radiation produces a perturbation of 1 part in 10^5 in the cosmic microwave background, this being the order of its non-cosmological in homogeneities as set by observations. Then if the overall density of solitons is equal to the critical density 1×10^{-29}g cm^{-3}, it turns out that each one has a mass of approximately 5×10^{11} that of the Sun. In other words, in order that dark solitons not perturb the microwave background too much, they must be less massive than a typical galaxy.

(b) Galaxy morphology can be used to bolster this inference. It is well known that objects consisting of conventional dark matter must be less than of order $10^8 M_\odot$ if the ones near to regular galaxies are not to affect the shapes of the latter through tidal interactions. A similar argument can be applied to solitons. However, it is not so straightforward, because by Eq. (5.5) the gravitational mass of a soliton varies depending on the relative influences of the 4D and extra parts of the metric. This implies that some solitons, including the kind considered by Gross and Perry [4], would not be detectable by their gravitational effects. In contrast, those with conventional masses must be less than about 10^8 of the mass of the Sun if they are located in the Local Group of galaxies.

(c) Gravitational lensing is a more clear-cut effect, since the deflection of a photon passing near a soliton is in general different from that with a conventional object. The relevant calculation was first carried out by Kalligas *et al.* in their analysis of the classical tests of relativity in the solar system ([30]; see also [31]). And while it is now widely acknowledged that the Sun's field is described by the canonical metric rather than the soliton metric, the latter can be applied to other astrophysical objects. The crux of the test is that the deflection angle of a photon by a soliton can be expressed as a combination of two terms: one is the conventional term of general relativity, and the other depends on the parameters of the soliton. The terms depend differently on the distance of recent approach, something which allows the separation of the influences due to solitons and conventional objects, at least in principle. Unfortunately, this is difficult in practice, especially in regard to microlensing. What is really needed for this test is accurate information on a well-defined system, such as the lensing of the light from a quasar by an isolated soliton.

The preceding three ways of detecting astrophysical solitons leave open the question of whether they make up a significant portion of the dark material in the universe. A comprehensive discussion of this topic is given in the book by Overduin and Wesson [45]. This also considers particle candidates. Most of these can be eliminated, especially if they are of conventional type and decay or annihilate to produce photons. However, it is worth mentioning the possibility that if solitons exist, they are microscopic as opposed to macroscopic in nature. It should be noted that there have been attempts to model elementary particles based on Einstein's equations, possibly in combination with Maxwell's equations, but none has met with widespread acceptance. For example, Bonnor and Cooperstock applied the junction conditions of general relativity to an electron situated in vacuum, and came to the conclusion that it is so small in relation to its mass that it must contain matter with a negative density [46]. This is rather unpalatable. However, Ponce de Leon has pointed out that a small mass can be obtained if the material is anisotropic with unequal radial and azimuthal pressures [47]. In fact, the effective mass approaches zero if the gravitational density $(\rho + 3p)$ tends to zero. While the conditions considered by Ponce de Leon are difficult to achieve in the 4D theory, they may be easier to approach in the 5D theory. Indeed, the pressures defined by Eqs. (5.3) are anisotropic and the gravitational mass (5.5) can be arbitrarily small. As to other effects, it is instructive to anticipate a result from the following chapter, which concerns the energy of charged solitons [20-23]. With the metric in quasi-Schwarzschild form and the consistency relation (5.2) between the two defining parameters the same as before, we can introduce a third parameter which defines the charge-to-mass ratio of the source: $k \equiv Q^2/4M^2(a-b)^2$. Then the metric-defined energy is given by

$$E = M(a + b/2) - Mk(a/2 + b) \ . \tag{5.15}$$

This includes the contributions to the energy of the gravitational, scalar and electromagnetic fields of a static soliton. As such, it could describe a charged dark-matter particle.

5.6. Conclusion

It is an understatement to say that solitons are strange objects. Their history is replete with rediscoveries and misidentifications (Section 5.2). The properties of static solitons are unusual, though they blend with those of their time-dependent counterparts (Sections 5.3 and 5.4; the electromagnetic properties of what might be called charged solitons are discussed in Chapter 6 following). Solitons represent predictions of the 5D theory as opposed to the usual 4D one, and there are in principle several ways of detecting them (Section 5.5). But perhaps their main importance is conceptual, in regard to what they tell us about the nature of gravitational and scalar fields.

The soliton metric in isotropic form (5.1) shows that the gravitational field of spacetime and the scalar field associated with the fifth coordinate are governed in strength by two dimensional parameters which are interdependent by Eq. (5.2). The 5D metric implies 4D matter with anisotropic pressure (5.3) and gravitational mass (5.5). Surprisingly, the latter goes to zero at the centre of coordinates. The metric takes a more familiar form in quasi-Schwarzschild coordinates (5.6). However, the Schwarzschild metric of 4D general relativity is more naturally embedded in the 5D canonical metric treated in Chapter 7. The 5D soliton metric has parameters in general which, via the manipulations (5.7) and (5.8), can be regarded as being dictated by a kind of moving angle. This shows that the gravitational and scalar fields arise from one source, which is bivalent.

Since there is technically nothing of a gravitational nature at the centre of a soliton, perhaps the best mental picture is an old-fashioned one, where lines of force emanate from a defect in the geometry.

References

[1] Wesson, P.S., *Five-Dimensional Physics* (World Scientific, Singapore, 2006).

[2] Kramer, D., *Acta. Phys. Pol.* **B2**, 807 (1970).

[3] Sorkin, R. D., *Phys. Rev. Lett.* **51**, 87 (1983).

[4] Gross, D.J. and Perry, M.J., *Nucl. Phys.* **B226**, 29 (1983).

[5] Davidson, A. and Owen, D.A., *Phys. Lett.* **B155**, 247 (1985).

[6] Dereli, T., *Phys. Lett.* **B161**, 307 (1985).

[7] Sokolowski, L. and Carr, B.J., *Phys. Lett.* **B176**, 334 (1986).

[8] Liu, H., *Gen. Rel. Grav.* **23**, 759 (1991).

[9] Agnese, A.G., Billyard, A.P., Liu, H. and Wesson, P.S., *Gen. Rel. Grav.* **31**, 527 (1999).

[10] Chatterjee, S., *Astron. Astrophys.* **230**, 1 (1990).

[11] Millward, R.S., arXiv:gr-qc/0603132 (2006).

[12] Wesson, P.S., *Phys. Lett.* **B276**, 299 (1992).

[13] Wesson, P.S., *Astrophys. J.* **394**, 19 (1992).

[14] Wesson, P.S., Ponce de Leon, J., *J. Math. Phys.* **33**, 3883 (1992).

[15] Wesson, P.S., *Space-Time-Matter,* 2nd. ed. (World Scientific, Singapore, 2007).

[16] Wesson, P.S., Ponce de Leon, J., *Class. Quantum Grav.* **11**, 1341 (1994).

[17] Bombelli, L., Koul, R.K., Kunstatter, G., Lee, J. and Sorkin, R.D., *Nucl. Phys.* **B289**, 735 (1987).

[18] Deser, S. and Soldate, M., *Nucl. Phys.* **B311**, 739 (1989).

[19] Billyard, A., Wesson, P.S. and Kalligas, D., *Int. J. Mod. Phys.* **A10**, 639 (1995).

[20] Sajko, W.N. and Wesson, P.S., *Gen. Rel. Grav.* **32**, 1381 (2000).

[21] Sajko, W.N. and Wesson, P.S., *Mod. Phys. Lett.* **A16**, 627 (2001).

[22] Liu, H. and Wesson, P.S., *Phys. Lett.* **B381**, 420 (1996).

[23] Liu, H. and Wesson, P.S., *Class. Quantum Grav.* **14**, 1651 (1997).

[24] Nodvik, J.S., *Phys. Rev. Lett.* **55**, 2519 (1985).

[25] Liko, T., *Phys. Lett.* **B617**, 193 (2005).

[26] Ponce de Leon, J., *Class. Quant. Grav.* **23**, 3043 (2006).

[27] Ponce de Leon, J., arXiv:gr-qc/0703094 (2007).

[28] Ponce de Leon, J., *Int. J. Mod. Phys.* **D17**, 237 (2008).

[29] Lake, K., *Class. Quant. Grav.* **23**, 5876 (2006).

[30] Kalligas, D., Wesson and P.S., Everitt, C.W.F., *Astrophys. J.* **439**, 548.

[31] Lim, P.H., Overduin, J.M. and Wesson, P.S., *J. Math. Phys.* **36**, 6907 (1995).

[32] Liu, H. and Wesson, P.S., *Class. Quant. Grav.* **13**, 2311 (1996).

[33] Liu, H. and Overduin, J.M., *Astrophys. J.* **538**, 386 (2000).

[34] Overduin, J.M., *Phys. Rev. D* **62**, 102001 (2000).

[35] Mashhoon, B., Liu H. and Wesson, P.S., *Phys. Lett.* **B331**, 305 (1994).

[36] Rippl, S., Romero, C. and Tavakol, R., *Class. Quant. Grav.* **12**, 2411 (1995).

[37] Romero, C., Tavakol, R. and Zalaletdinov, R., *Gen. Rel. Grav.* **28**, 365 (1996).

[38] Lidsey, J.E., Romero, C., Tavakol R. and Rippl, S., *Class. Quant. Grav.* **14**, 865 (1997).

[39] Seahra, S.S. and Wesson, P.S., *Class. Quant. Grav.* **20**, 1321 (2003); Wesson, P.S., arXiv:1104.3244 (2011).

[40] The main text uses a, b for the two dimensionless parameters which determine the soliton class of metrics. These relate to the terminology used by previous workers as follows: $a = A - B, b = 2B$ (Kramer, Ref. [2]); $a = 1/\alpha, b = \beta/\alpha$ (Gross and Perry, Ref. [4]), though their α, β should not be confused with those used in Eq. (5.8) of the main text); $a = \varepsilon\kappa, b = -\varepsilon$ (Davidson and Owen, Ref. [5]).

[41] Liu, H., Wesson, P.S. and Ponce de Leon, J., *J. Math Phys.* **34**, 4070 (1993).

[42] Ponce de Leon, J. and Wesson, P.S., *J. Math. Phys.* **34**, 4080 (1993).

[43] Wesson, P.S., Liu, H. and Lim, P., *Phys. Lett.* **B298**, 69 (1993).

[44] Wesson, P.S., *Astrophys. J* **420**, L49 (1994).

[45] Overduin, J.M. and Wesson, P.S., *The Light/Dark Universe* (World Scientific, Singapore, 2008).

[46] Bonnor, W.B. and Cooperstock, F.I., *Phys. Lett.* **A139**, 442 (1989).

[47] Ponce de Leon, J., *Gen. Rel. Grav.* **36**, 1453 (2004); see also Ponce de Leon, J., arXiv:0207108 (2002).

Chapter 6

Electromagnetism in 5D

6.1. Introduction

In 1921, Kaluza showed that if the world were extended to five dimensions, the resulting field equations (if independent of the extra coordinate) contained the vacuum Einstein and Maxwell equations. And in 1926, Klein suggested that if the fifth dimension was rolled up or compactified to a microscopic size, the momentum associated with it would explain the quantization of the electron charge. More recently, various workers have shown that Space-Time-Matter theory as outlined in Chapter 2 can explain the phenomenological properties of matter if the traditional geometry of Kaluza-Klein theory is broadened to include dependence on the extra coordinate and the extra dimension is allowed to be 'large'. However, it is true that in this approach photons still occupy a special place. In the present chapter, we wish to see how electromagnetism fits into a general 5D geometry.

Thus we start with a 5D metric that depends on $x^4 = \ell$ and redo the algebra that puts it into a form where we recognize the electromagnetic potentials (Section 6.2). Then we work out the fully covariant form of the 5D geodesic equation (Section 6.3). We will find that the latter now contains x^4-dependent terms which are in principle important for testing 5D field theory. These terms are small in practice, however, so subsequently we drop x^4-dependence and present a class of exact solutions in 5D which extends the soliton class by including electric charge (Section 6.4). This class includes the case where the 4D part of the metric describes a Schwarzschild-like black hole. To elucidate the effects of charge in 5D we then give an analysis of the electrodynamics involved in this case, finding that it leads to unexpected differences from 4D (Section 6.5). Next, we return to the field equations and the question of induced matter, and present the former in a way that allows us to evaluate the energy density of the electromagnetic field and the energy of charged solitons (Section 6.6). Finally we carry out an analysis to obtain explicit expressions for the charge and mass of a particle in 5D (Section 6.7). We will find that while 5D electromagnetism has been much studied, it still contains some surprises.

6.2. Metrics and Potentials

The reduction of a 5D metric to a 4D one where the potentials can be recognized as those of gravity and electromagnetism has been done by various workers under the assumption of x^4-independence. In the case where there is dependence on x^4, the algebra becomes significantly more complicated. There is also controversy of a conceptual kind, which goes back to the fact that the field equations are now 5D in nature. This means that, in principle, coordinate transformations which involve $x^4 = \ell$ can alter the form of 4D relations. This is objectionable to some workers, who prefer in practice to maintain the invariance of 4D relations by altering the 5D equations of motion. However, this difference of opinion affects mainly the definition of a *neutral* test mass; so we defer a discussion of this issue to elsewhere, and here follow the analysis of Wesson and Ponce de Leon for a charged particle [3], making the steps in the algebra explicit.

The 5D interval is given by

$$dS^2 = \gamma_{AB} dx^A dx^B \; .$$
(6.1)

If we wish to investigate the 4D interval ds included in this, it is convenient to define a unit 5D vector Ψ^A, tangent to the fifth dimension:

$$\Psi^A \equiv \frac{\delta_4^A}{\sqrt{\varepsilon \gamma_{44}}} \quad . \tag{6.2}$$

This will enable us to split the 5D metric into a part parallel to Ψ^A and a 4D part orthogonal to it. Thus we define a projector:

$$g_{AB} \equiv \gamma_{AB} - \varepsilon \Psi_A \Psi_B \quad , \tag{6.3}$$

where by Eq. (6.2)

$$g_{AB} = \gamma_{AB} - \frac{\gamma_{4A}\gamma_{4B}}{\gamma_{44}} \quad . \tag{6.4}$$

This has

$$g_{44} = g_{4A} = 0 \quad , \tag{6.5}$$

so the 5D line element (6.1) becomes the sum of a 4D line element and an extra part:

$$dS^2 = g_{\mu\nu}dx^\mu dx^\nu + \gamma_{44}\left(dx^4 + \frac{\gamma_{4\alpha}}{\gamma_{44}}dx^\alpha\right)^2 \quad , \tag{6.6}$$

$$ds^2 \equiv g_{\mu\nu}dx^\mu dx^\nu \quad . \tag{6.7}$$

Here all of $g_{\mu\nu}, \gamma_{44}$ and $\gamma_{4\alpha}$ can depend on x^4 and the signature is open via the choice $\varepsilon = \pm 1$ for this parameter. To see what is involved here, let us introduce a 4-vector and a scalar defined by

$$A_\mu \equiv \frac{\gamma_{\mu 4}}{\gamma_{44}} \quad , \tag{6.8}$$

$$\Phi^2 \equiv \varepsilon \gamma_{44} \quad . \tag{6.9}$$

With these, Eqs. (6.6) and (6.7) give

$$dS^2 = ds^2 + \varepsilon \Phi^2 \left(dx^4 + A_\mu dx^\mu\right)^2 \quad . \tag{6.10}$$

The 5D metric and the 4D metric are of course associated with groups of coordinate transformations which, as we noted before, are of the form

$$x^A \to \bar{x}^A(x^B) \ , \tag{6.11}$$

$$x^\mu \to \bar{x}^\mu(x^\nu) \ . \tag{6.12}$$

Since the group (6.11) is broader than (6.12), the geodesics will have extra terms and may have unfamiliar forms. (In this chapter we will only treat geodesics with $dS^2 \neq 0$.) However, it will still be the case that $g_{\mu\nu}$ is associated with gravity, A_μ with electromagnetism, and Φ with a scalar interaction.

We will need some further relations in what follows. Eq. (6.4) with $A = \alpha$ and $B = \beta$ when combined with Eqs. (6.8) and (6.9) gives

$$\gamma_{\alpha\beta} = g_{\alpha\beta} + \varepsilon\Phi^2 A_\alpha A_\beta \ . \tag{6.13}$$

While if we take $\gamma^{AC}\gamma_{BC} = \delta^A_B$ with $A = \mu$, $B = 4$ and $C = \lambda, 4$ we obtain $\gamma^{\mu4} = -\gamma^{\mu\lambda}\gamma_{4\lambda}/\gamma_{44}$. This can be used with Eq. (6.4) in the expansion of $\gamma^{AC}\gamma_{BC} = \delta^A_B$ with $A = \mu$, $B = \nu$ and $C = \lambda, 4$ to find

$$\gamma^{\mu\nu} = g^{\mu\nu} \ . \tag{6.14}$$

Also, since $\gamma^{\mu4} = -\gamma^{\mu\lambda}A_\lambda$ we see that

$$\gamma^{4\mu} = -A^\mu \ . \tag{6.15}$$

And from Eqs. (6.8) and (6.9) we have

$$\gamma_{4\mu} = \varepsilon\Phi^2 A_\mu \ . \tag{6.16}$$

These relations will be useful, especially Eq. (6.14) which will allow us to raise indices using either the 5D or the 4D metric.

6.3. Geodesics

By extremizing the interval S we obtain the 5D geodesic equation

$$\frac{d^2x^A}{dS^2} + \Gamma^A_{BC} \frac{dx^B}{dS} \frac{dx^C}{dS} = 0 \ , \tag{6.17}$$

where the Christoffel symbol of the second kind is

$$\Gamma^A_{BC} = \frac{\gamma^{AD}}{2} \left(\frac{\partial \gamma_{BD}}{\partial x^C} + \frac{\partial \gamma_{CD}}{\partial x^B} - \frac{\partial \gamma_{BC}}{\partial x^D} \right) . \tag{6.18}$$

It is convenient to consider the five components of Eq. (6.17) in the order 4,123,0, as follows.

The $A = 4$ component of Eq. (6.17) can be rewritten as

$$\gamma_{4A} \frac{d^2x^A}{dS^2} + \gamma_{4A} \Gamma^A_{CD} \frac{dx^C}{dS} \frac{dx^D}{dS} = 0 \ . \tag{6.19}$$

It may be verified by expansion, using Eqs. (6.8) and (6.9), that Eq. (6.19) is equivalent to

$$\frac{dn}{dS} = \frac{1}{2} \frac{\partial \gamma_{CD}}{\partial x^4} \frac{dx^C}{dS} \frac{dx^D}{dS} \ , \tag{6.20}$$

where n is a scalar function:

$$n \equiv \varepsilon \Phi^2 \left(\frac{dx^4}{dS} + A_\alpha \frac{dx^\alpha}{dS} \right) . \tag{6.21}$$

From Eq. (6.20) we see that if the 5D metric γ_{CD} were independent of x^4, then n would be a constant of the motion. [This in the sense that $dn/dS = (\partial n/\partial x^A)(dx^A/dS) = 0$ even though n could depend on x^A.] However, this will not be the case in general. We also note from Eqs. (6.10) and (6.21) that

$$dS = \frac{ds}{(1-\varepsilon n^2/\Phi^2)^{1/2}} \ , \tag{6.22}$$

which gives the 5D interval in terms of its 4D part.

The $A = 123$ components of Eq. (6.17) give the equations of motion of a test particle in ordinary space. However, to obtain these explicitly we need to evaluate the following components of Eq. (6.18) using Eqs. (6.13)–(6.16) above:

$$\Gamma^{\mu}_{44} = -\varepsilon\Phi\nabla^{\mu}\Phi + \varepsilon\Phi\left(A^{\mu}\frac{\partial\Phi}{\partial x^4} + \Phi g^{\mu\lambda}\frac{\partial A_{\lambda}}{\partial x^4}\right) , \tag{6.23}$$

$$\Gamma^{\mu}_{4v} = -\varepsilon\Phi\nabla^{\mu}\Phi A_v - \frac{\varepsilon\Phi^2}{2}F^{\mu}_{\ v} + \frac{g^{\mu\lambda}}{2}\frac{\partial\gamma_{\lambda v}}{\partial x^4} , \tag{6.24}$$

$$\hat{\Gamma}^{\mu}_{\alpha\beta} = \Gamma^{\mu}_{\alpha\beta} - \varepsilon\Phi\nabla^{\mu}\Phi A_{\alpha}A_{\beta} - \frac{\varepsilon\Phi^2}{2}\left(A_{\alpha}F^{\mu}_{\ \beta} + A_{\beta}F^{\mu}_{\ \alpha}\right)$$
$$+ \frac{A^{\mu}}{2}\frac{\partial\gamma_{\alpha\beta}}{\partial x^4} . \tag{6.25}$$

Here $\nabla^{\mu}\Phi = g^{\mu v}\nabla_v\Phi = g^{\mu v}(\partial\Phi/\partial x^v)$ involves the usual 4D covariant derivative, $\Gamma^{\mu}_{\alpha\beta}$ is the usual 4D Christoffel symbol defined by a relation analogous to Eq. (6.18), and $F_{\mu v}$ is the usual 4D antisymmetric tensor

$$F_{\mu v} \equiv \frac{\partial A_v}{\partial x^{\mu}} - \frac{\partial A_{\mu}}{\partial x^v} = \nabla_{\mu}A_v - \nabla_v A_{\mu} . \tag{6.26}$$

To bring out both the gravitational and electromagnetic parts of the geodesic (6.17) in a recognizable form, we need to substitute for Eqs. (6.23)–(6.26) and use the definition (6.21) for the quantity n. After considerable algebra, we find that Eq. (6.17) with $A = \mu$ can be written

$$\frac{d^2x^{\mu}}{ds^2} + \Gamma^{\mu}_{\alpha\beta}\frac{dx^{\alpha}}{ds}\frac{dx^{\beta}}{ds} = nF^{\mu}_{\ \alpha}\frac{dx^{\alpha}}{ds} + \varepsilon n^2\frac{\nabla^{\mu}\Phi}{\Phi^3} - A^{\mu}\frac{dn}{ds}$$
$$- g^{\mu\lambda}\frac{dx^4}{ds}\left(n\frac{\partial A_{\lambda}}{\partial x^4} + \frac{\partial g_{\lambda v}}{\partial x^4}\frac{dx^v}{ds}\right) . \tag{6.27}$$

If we replace total derivatives with respect to S by their counterparts with respect to s using Eq. (6.22), we obtain

$$\frac{d^2x^{\mu}}{ds^2} + \Gamma^{\mu}_{\alpha\beta}\frac{dx^{\alpha}}{ds}\frac{dx^{\beta}}{ds} = \frac{n}{(1-\varepsilon n^2/\Phi^2)^{1/2}}$$
$$\times\left(F^{\mu}_{\ v}\frac{dx^v}{ds} - \frac{A^{\mu}}{n}\frac{dn}{ds} - g^{\mu\lambda}\frac{\partial A_{\lambda}}{\partial x^4}\frac{dx^4}{ds}\right)$$
$$+ \frac{\varepsilon n^2}{(1-\varepsilon n^2/\Phi^2)\Phi^3}\left[\nabla^{\mu}\Phi + \left(\frac{\Phi}{n}\frac{dn}{ds} - \frac{d\Phi}{ds}\right)\frac{dx^{\mu}}{ds}\right]$$
$$- g^{\mu\lambda}\frac{\partial g_{\lambda v}}{\partial x^4}\frac{dx^v}{ds}\frac{dx^4}{ds} . \tag{6.28}$$

This is the fully general equation of motion in Kaluza-Klein theory. We will look at the $\mu = 0$ component which is included in Eq. (6.28) below, but the $\mu = 123$ components deserve comment now. Thus, the left-hand side of Eq. (6.28), if zero, would define geodesic motion in 4D Einstein gravity, but the terms on the right-hand side modify the motion. Of these, the three inside the first bracket relate to electromagnetism. All three vanish if the electromagnetic potentials A_μ are zero; i.e., if $\gamma_{4\mu} = 0$ by Eq. (6.8). Of these three terms, the last two would vanish independently if the metric did not depend on x^4 (since then n would be a constant of the motion by the above so $dn/ds = 0$). This leaves the first term, which we recognize as the usual Lorentz force provided we write

$$\frac{q}{m} = \frac{n}{(1-\varepsilon n^2/\Phi^2)^{1/2}}$$

(6.29)

for the charge-to-mass ratio of the test particle.

The $A = 0$ component of Eq. (6.17) can usefully be treated separately from the preceding discussion (where it is included implicitly). This component can be written

$$\gamma_{0A}\frac{d^2x^A}{ds^2} + \Gamma_{0,CD}\frac{dx^C}{ds}\frac{dx^D}{ds} = 0 \ .$$

(6.30)

It may be verified by expansion, using Eqs. (6.13)–(6.16) and (6.21), that Eq. (6.30) is equivalent to

$$\frac{d}{ds}\left(g_{00}\frac{dx^0}{ds} + g_{0i}\frac{dx^i}{ds}\right) = -\frac{d(nA_0)}{ds} + \frac{1}{2}\frac{\partial\gamma_{CD}}{\partial x^0}\frac{dx^C}{ds}\frac{dx^D}{ds} \ .$$

(6.31)

Here $i = 123$. If as before we replace the total derivatives with respect to S by their counterparts with respect to s using Eq. (6.22), in all but the last factor (which we leave mixed for algebraic convenience), we obtain

$$\frac{d}{ds}\left[g_{00}\left(1 - \frac{\varepsilon n^2}{\Phi^2}\right)^{\frac{1}{2}}\left(\frac{dx^0}{ds} + \frac{g_{0i}}{g_{00}}\frac{dx^i}{ds}\right) + nA_0\right] =$$
$$\frac{1}{2}\frac{\partial\gamma_{CD}}{\partial x^0}\frac{dx^C}{ds}\frac{dx^D}{ds} \ .$$

(6.32)

This relation looks complicated. But it can be simplified greatly by using an argument analogous to what was done above in Eqs. (6.2)–(6.7) and (6.19)–(6.21). Thus we define a unit 4D vector τ^μ tangent to the time coordinate:

$$\tau^\mu \equiv \frac{\delta^\mu_0}{\sqrt{g_{00}}} \ . \tag{6.33}$$

We also define a projector onto the 3-space orthogonal to τ^μ as:

$$\lambda_{\mu\nu} \equiv \tau_\mu \tau_\nu - g_{\mu\nu} = \frac{g_{\mu 0} g_{\nu 0}}{g_{00}} - g_{\mu\nu} \ . \tag{6.34}$$

Then $\lambda_{00} = \lambda_{0j} = 0$ analogous to Eq. (6.5), and the 4D line element (6.7) becomes the sum of a time part and a space part:

$$ds^2 = g_{00}\left(dx^0 + \frac{g_{0i}}{g_{00}}dx^i\right)^2 - \lambda_{ij}dx^i dx^j \ . \tag{6.35}$$

This can be rewritten and factored to give

$$\frac{dx^0}{ds} = \frac{g_{0j}}{g_{00}}\frac{dx^j}{ds} = \frac{1}{(1-v^2)^{1/2}(g_{00})^{1/2}} \ , \tag{6.36}$$

provided the square of the spatial 3-velocity is defined by $v^2 \equiv \lambda_{ij}v^i v^j$ and the velocity itself is defined by

$$v^i \equiv \frac{dx^i}{(g_{00})^{1/2}[dx^0 + (g_{0i}/g_{00})dx^i]} \ . \tag{6.37}$$

This is an element of ordinary space divided by an element of proper time in Eq. (6.35), and so is the appropriate definition of a 3-velocity analogous to the usual one for a 4-velocity. We can now use Eq. (6.36) to substitute for the term in square brackets in Eq. (6.32). The result is

$$\frac{de}{ds} = \frac{1}{2}\frac{\partial \gamma_{CD}}{\partial x^0}\frac{dx^C}{ds}\frac{dx^D}{ds} \ , \tag{6.38}$$

where e is a scalar function:

$$e \equiv \frac{(g_{00})^{1/2}}{(1-v^2)^{1/2}} \left(1 - \frac{\varepsilon n^2}{\Phi^2} \right)^{1/2} + nA_0 \quad . \tag{6.39}$$

The relations (6.38) and (6.39) are equivalent to the geodesic (6.30). From Eq. (6.38) we see that if the 5D metric γ_{CD} were independent of x^0, then e would be a constant of the motion. However, this will not be the case in general. We do note, though, that Eq. (6.39) bears a strong resemblance to the expression for the energy $\in$ of a test particle in a static field in 4D general relativity as given by Landau and Lifshitz [4]:

$$\in = \frac{(g_{00})^{1/2}}{(1-v^2)^{1/2}} m + qA_0 \quad . \tag{6.40}$$

A comparison together with previous relations shows that the mass of a test particle is related to the dependence of the metric on the extra coordinate, while its electric charge is related to its rate of motion in the extra dimension.

Explicit expressions for the charge and mass of a particle in 5D will be derived in Section 6.7, using a somewhat different method and an alternative form for the metric. The analysis carried out above follows what might be termed the traditional approach, in which a general metric (6.1) is put into the usual electromagnetic form (6.10) and used to evaluate the components of the 5D geodesic equation, where Eq. (6.20) is for the fifth dimension and Eq. (6.28) for spacetime. Our equations of motion include effects caused by the dependence of the metric on $x^4 = \ell$, but reduce to others in the literature when this is absent [5-9]. An important case where there is ℓ-dependence is canonical space [10, 11]. We will touch on this case in Section 6.7, but it is not treated fully until Chapter 7. There are, however, several situations in which the metric does not depend on ℓ but there are interesting physical effects. Prime among these are the solitons of the preceding chapter, and we now inquire about the properties of these when they have electric charge.

6.4. Charged Solitons and Black Holes

The initial application of 5D to electromagnetism was motivated, as we saw, by Kaluza's discovery that the empty 5D field equations contained the corresponding 4D ones of Einstein and Maxwell. Since that time, the standard metric for 5D electromagnetism has been Eq. (6.10) considered above. That metric has many applications [12, 13]. However, potentially greater insight into the effects of electric charge in 5D may be obtained by considering the solitons, which have only recently been fully understood. A summary of the properties of neutral solitons was given in Chapter 5; now we wish to follow Liu and Wesson [14, 15] and examine charged solitons.

The metric and solution are given by:

$$dS^2 = \frac{(1-k)B^a}{(1-kB^{a-b})} dt^2 - B^{-a-b} dr^2 - r^2 B^{1-a-b} d\Omega^2$$
$$- \frac{(B^b - kB^a)}{(1-k)} (d\ell + A dt)^2 \quad ,$$

$$A \equiv \frac{-\sqrt{k}(1 - B^{a-b})}{(1 - kB^{a-b})} \quad , \tag{6.41}$$

$$B \equiv 1 - \frac{2(1-k)M}{r} \quad ,$$

$$1 = a^2 + b^2 + ab \quad .$$

Here k is a dimensionless parameter whose nature we will elucidate below; A and B are dimensionless potentials that depend on k, r and the mass M at the centre of the 3-geometry; and a, b are dimensionless constants that obey the same consistency relation as before. The choice $a = 1, b = 0, k = 0$ makes the 4D part of the 5D metric Schwarzschild. The choice $a = 1, b = -1, k = 0$ makes the diagonal part of the 5D metric take the form considered by Chatterjee [16]. The limit $k \to 0$ in general makes the off-diagonal part of the metric disappear and gives back the Gross-Perry/Davidson-Owen class of solitons considered in Section 5.3. The latter is a 2-parameter class, but Eq. (6.41) is a 3-parameter class, because it depends on k, M and one or the other of a, b. Clearly, we need to identify the new parameter k.

To do this, let us examine the case $a = 1, b = 0, k \neq 0$. Unlike the other solitons (which do not possess event horizons of the conventional sort), there is some justification for referring to this object as a black hole, because for $k \to 0$ the metric (6.41) reduces exactly to Schwarzschild plus an extra flat piece. In general, the metric can now be written in the following form:

$$dS^2 = BE^{-1}dt^2 - B^{-1}dr^2 - r^2d\Omega^2 - E(d\ell + Adt)^2 \quad,$$

$$E \equiv \frac{(1-kB)}{(1-k)} = 1 + \frac{2Mk}{r} \quad,$$

$$A \equiv \frac{-\sqrt{k}(1-B)}{(1-kB)} = -\frac{2M\sqrt{k}}{Er} \quad,$$

$$B \equiv 1 - \frac{2M(1-k)}{r} = E - \frac{2M}{r} \quad.$$

(6.42)

This is a 2-parameter (k, M) class, where A, B, E are the potentials. Of these, A is actually the electrostatic potential, as may be confirmed by comparing Eq. (6.42) with the Kaluza-Klein metric written in the standard form where the field equations yield Maxwell's equations. The electric field has only a radial component (E_1, not to be confused with E above). In terms of the usual Faraday tensor $F_{\alpha\beta} \equiv \partial A_\beta / \partial x^\alpha - \partial A_\alpha / \partial x^\beta$ of Eq. (6.26), it is given by

$$E_1 = F_{10} = \frac{\partial A}{\partial r} = \frac{2M\sqrt{k}}{E^2r^2} \quad.$$

(6.43)

The associated invariant is

$$I \equiv F^{\alpha\beta}F_{\alpha\beta} = -2E(F_{10})^2 = -\frac{8M^2k}{E^3r^4} \quad.$$

(6.44)

By Eqs. (6.42) for $r \to \infty$, we have $B \to 1, E \to 1$ and $A \to -2M\sqrt{k}/r$. The last should be $-Q/r$ to agree with Coulomb's law at large distances from the source charge Q. This enables us to identify the dimensionless parameter k with conventional units restored as

$$k = \frac{Q^2}{4GM^2} .$$
(6.45)

Using this in Eqs. (6.42)–(6.44) allows us to sum up the quantities of interest (valid for any r) as follows:

$$E = 1 + \frac{Q^2}{2Mr} ,$$

$$A = -\frac{Q}{Er} , \quad B = E - \frac{2M}{r} ,$$
(6.46)

$$E_1 = \frac{Q}{E^2 r^2} , \quad I = -\frac{2Q^2}{E^3 r^4} .$$

Already here we can see that a quantity Q^2/Mr that depends on the *combination* of Q and M comes into the 5D solution, whereas only the quantities M/r and Q^2/r^2 that depend *separately* on Q and M come into the corresponding 4D (Reissner-Nordström) solution.

The preceding paragraph concerns the case $a = 1, b = 0, k \neq 0$. However, the choice $a = 1, b = 0$ is smoothly contained in the range $-2/\sqrt{3} < (a, b) < 2/\sqrt{3}$ allowed by the consistency relation given by the last member of Eqs. (6.41). Thus we identify the whole 3-parameter class of exact solutions (6.41) as charged solitons.

6.5. Dynamics of Charged Black Holes

In this section, we will combine what we learned about geodesic motion in Section 6.3 with the solution presented in Section 6.4 to study the dynamics of a test particle with charge q and mass m in orbit around a 5D black hole with charge Q and mass M.

The metric for the latter is Eq. (6.42), which is independent of x^4 so n in Eq. (6.21) is a constant of the motion. That relation is then the first integral of the last component of the geodesic equation. The spacetime components are given by Eq. (6.28). The equations of motion are thus

$$\frac{d\ell}{ds} + A\frac{dt}{ds} = -\frac{n}{E} \quad,$$

(6.47)

$$\frac{d^2x^\alpha}{ds^2} + \Gamma^\alpha_{\beta\gamma}\frac{dx^\beta}{ds}\frac{dx^\gamma}{ds} = \frac{n}{(1+n^2/E)^{1/2}}F^\alpha{}_\beta\frac{dx^\beta}{ds}$$
$$- \frac{n^2}{(1+n^2/E)}\frac{\partial_\beta E}{2E^2}\left(g^{\alpha\beta} - \frac{dx^\alpha}{ds}\frac{dx^\beta}{ds}\right).$$

(6.48)

By Eqs. (6.29) or (6.48) directly, we obtain the asymptotic $(r \to \infty, E \to 1)$ value for the charge-to-mass ratio of the test particle as

$$\frac{q}{m} = \frac{n}{(1+n^2)^{1/2}} \quad, \quad \text{so } n = \frac{q}{(m^2 - q^2)^{1/2}} \quad.$$

(6.49)

The second term on the right-hand side of Eq. (6.48) is an extra force not present in 4D electrodynamics. However, it is tiny for classical as opposed to quantum situations because it depends on $(q/m)^2$, which with conventional units restored is $q^2/Gm^2 \ll 1$. For any r, we can rewrite Eqs. (6.48) and (6.49) as

$$\frac{d^2x^\alpha}{ds^2} + \Gamma^\alpha_{\beta\gamma}\frac{dx^\beta}{ds}\frac{dx^\gamma}{ds} = \left(\frac{q}{mD}\right)F^\alpha{}_\beta\frac{dx^\beta}{ds}$$
$$- \left(\frac{q}{mD}\right)^2\frac{\partial_\beta E}{2E^2}\left(g^{\alpha\beta} - \frac{dx^\alpha}{ds}\frac{dx^\beta}{ds}\right),$$

(6.50)

$$D \equiv \left(1 - \frac{e^2Q^2}{2m^2 MEr}\right)^{1/2},$$

where $E = (1 + Q^2/2Mr)$ as before. We can evaluate this for the index $\alpha = 1$ in what might be called the standard approximation, specified by $|q/m| \ll 1$, $M/r \ll 1$, $u^{123} \ll u^0 \approx (g^{00})^{1/2}$. In this approximation, $\Gamma^1_{00} \approx M/r^2$, $F_0{}^1 \approx Q/r^2$ and we obtain

$$m\frac{d^2r}{ds^2} = -\frac{mM}{r^2} + \frac{qQ}{r^2} \quad,$$

(6.51)

which is of course true in many situations.

Eqs. (6.47)–(6.49) can be studied analytically for two illustrative cases: (i) $Q = 0, q \neq 0$ and (ii) $Q \neq 0, q = 0$. For case (i) we have $k = 0$

by Eq. (6.45) and $E = 1$, $A = 0$ and $B = 1 - 2M/r$ from Eqs. (6.42). The metric in Eq. (6.42) is

$$dS^2 = \left(1 - \frac{2M}{r}\right) dt^2 - \frac{dr^2}{(1-2M/r)} - r^2 d\Omega^2 - d\ell^2 \quad , \qquad (6.52)$$

which is just a neutral 4D Schwarzschild black hole plus an extra flat dimension. The motion of a charged particle around such a black hole, from Eqs. (6.47)–(6.49), is given by

$$\frac{d\ell}{dS} = -n \quad , \quad n = \frac{q/m}{(1-q^2/m^2)^{1/2}} \quad , \qquad (6.53)$$

$$\frac{d^2x^\alpha}{ds^2} + \Gamma^\alpha_{\beta\gamma} \frac{dx^\beta}{ds} \frac{dx^\gamma}{ds} = 0 \quad . \qquad (6.54)$$

In other words, for this case the constant of the motion associated with the extra dimension is essentially the charge-to-mass ratio of the test particle, and the 4D motion is geodesic in the Einstein sense.

For case (ii), we have $n = 0$ by Eq. (6.49), so Eqs. (6.46)–(6.48) lead to

$$\frac{d\ell}{dS} + A \frac{dt}{dS} = 0 \quad , \quad A = -\frac{Q}{r}\left(1 + \frac{Q^2}{2Mr}\right)^{-1} \quad , \qquad (6.55)$$

$$\frac{d^2x^\alpha}{ds^2} + \Gamma^\alpha_{\beta\gamma} \frac{dx^\beta}{ds} \frac{dx^\gamma}{ds} = 0 \quad . \qquad (6.56)$$

The 4D motion is again geodesic in the Einstein sense, while Eq. (6.55) implies that the metric (6.42) loses its extra part and becomes effectively 4D. Using Eq. (6.46), we find that in the large-r approximation $B \approx (1 - 2M/r + Q^2/2Mr)$ and $BE^{-1} \approx (1 - 2M/r + Q^2/r^2)$. Thus the metric reads

$$ds^2 \approx \left(1 - \frac{2M}{r} + \frac{Q^2}{r^2}\right) dt^2 - \frac{dr^2}{(1-2M/r + Q^2/2Mr)} - r^2 d\Omega^2 \quad . \quad (6.57)$$

This can be compared to the Reissner-Nordström solution of Einstein's theory:

$$ds^2 = \left(1 - \frac{2M}{r} + \frac{Q^2}{r^2}\right) dt^2 - \frac{dr^2}{(1 - 2M/r + Q^2/r^2)} - r^2 \, d\Omega^2 \ . \quad (6.58)$$

We see that they agree in g_{00} but not in g_{11}. That is, at large distances the 3-space associated with the '5D' black hole (6.57) is different from that associated with the 4D black hole defined by Eq. (6.58). The difference will be minor in practice, since the mass equivalent of the energy associated with the central charge will be small in most situations $(GQ^2/c^4 r^2 \ll 1)$. However, this discrepancy prompts us to ask about the consistency of the electromagnetic field associated with the solutions (6.42). To be precise: The solutions (6.58) satisfy the 4D field equations $G_{\alpha\beta} = 8\pi T_{\alpha\beta}$ in terms of the Einstein tensor and a standard energy-momentum tensor describing an electrostatic field, whereas Eqs. (6.57) or (6.42) satisfy 5D field equations $R_{AB} = 0$ with nothing on the right-hand side; so what is the effective energy-momentum tensor of the 5D solutions (6.42) if they are viewed in 4D?

The answer to this can be given here by anticipating results from the following section on the induced matter associated with non-diagonal metrics such as Eq. (6.10). For the charged black-hole solutions defined by Eqs. (6.42), we find that the energy density associated with the electromagnetic field is given by

$$T_{00}^{\text{em}} = \frac{BM^2 k}{E^3 r^4} \rightarrow \frac{Q^2}{4r^4} \quad (r \rightarrow \infty) \ . \quad (6.59)$$

This has its conventional form, implying that the 5D solutions (6.42) do not show departures from the 4D ones in the far field, even though they may in the near field.

The above concerns the electromagnetic field. Returning to dynamics, we expect small differences in the orbits of particles about the central mass, because the Lagrangian associated with the 5D charged black hole (6.42) differs from that associated with the 4D charged black hole (6.58) and also from that associated with the 5D neutral soliton (5.6). The latter was investigated extensively in the preceding chapter, and it is instructive to see how things are altered by the presence of electric charge. To examine this, let us rewrite Eqs. (6.42) to give the Lagrangian

$$\mathcal{L} = BE^{-1}\dot{t}^2 - B^{-1}\dot{r}^2 - r^2(\dot{\theta}^2 + \sin^2\theta\dot{\phi}^2)$$
$$-E(\dot{\ell} + A\dot{t})^2 \ . \tag{6.60}$$

Here $\dot{t} = dt/d\lambda$ etc., where λ is an affine parameter, and we can assume without loss of generality that the orbit is confined to the plane $\theta = \pi/2$ with $\dot{\theta} = 0$. Then from Eq. (6.60) we obtain three constants of the motion in the usual way:

$$j \equiv BE^{-1}\dot{t} - E(\dot{\ell} + A\dot{t})A \ ,$$

$$h \equiv r^2\dot{\phi} \ , \tag{6.61}$$

$$n \equiv -E(\dot{\ell} + A\dot{t}) \ .$$

Rearranging these gives the 5-velocities for the case where $\lambda = S$ (i.e., $\mathcal{L} = 1$) as

$$\dot{t} = EB^{-1}(j - nA) \ ,$$

$$\dot{\phi} = hr^{-2} \ , \tag{6.62}$$

$$\dot{\ell} = -nE^{-1} - EAB^{-1}(j - nA) \ .$$

For $Q = 0$ ($E = 1, A = 0$) these relations give $\dot{t} = B^{-1}j, \dot{\phi} = hr^{-2}$ and $\dot{\ell} = -n$. These results are identical to those of on older study which evaluated the classical tests of the solar system for the neutral soliton with parameters $a = 1, b = 0$ [17]. That application may well be considered obsolete, given the subsequent discovery of the canonical embedding for the Schwarzschild solution, which preserves 4D dynamics exactly (see Chapter 7). However we proceed with the analysis, because the isolation of *any* 5D test is significant.

The test we are considering concerns the shape of the orbit of a charged test particle around a source. (I.e., the 5D analog of the classical 4D test involving the perihelion advance of a planet.) To investigate this, we can take the Lagrangian (6.60) [with $\lambda = S$], substitute into it for the constants of motion (6.61) [with $Q, A \neq 0$ for now] and rearrange to get:

$$\frac{1}{B}\left(\frac{dr}{d\phi}\right)^2 = \frac{r^4}{h^2}\left[\frac{E}{B}(j - nA)^2 - \frac{n^2}{E} - 1\right] - r^2 \quad . \tag{6.63a}$$

Changing to a variable $u = 1/r$ as in classical mechanics, this reads

$$\left(\frac{du}{d\phi}\right)^2 = \frac{1}{h^2}\left[E(j - nA)^2 - \frac{n^2 B}{E} - B\right] - Bu^2 \quad . \tag{6.63b}$$

This formula could provide the basis for a detailed investigation of the orbit of a charged test particle about a charged source, but for now we look at the simpler case $Q = 0 (q \neq 0)$, for which Eq. (6.63b) gives

$$\left(\frac{du}{d\phi}\right)^2 = \frac{(j^2 - n^2 - 1)}{h^2} + \frac{(n^2 + 1)2Mu}{h^2} + 2Mu^3 - u^2 \quad . \tag{6.64}$$

Differentiating with respect to ϕ, and assuming that the orbit is not perfectly circular $(du/d\phi \neq 0)$, we obtain

$$\frac{d^2 u}{d\phi^2} = \frac{(1 + n^2)}{h^2}M + 3Mu^2 - u \quad . \tag{6.65}$$

This is a special case of the equation in the corresponding astrophysical problem [17]. The test particle has a precession per orbit which in our notation is

$$\delta\phi = \frac{6\pi M^2}{h^2}\left[1 + \frac{q^2}{(m^2 - q^2)}\right] \quad . \tag{6.66}$$

Here, the first term is the one quoted in the usual approach using Einstein's theory, while the second is an extra one that is necessarily positive. It exists because the Lagrangian (6.60) or metric (6.42) looks like the usual 4D one, but augmented by a boost in the extra dimension [$\dot{\ell} = d\ell/dS = -n = -q(m^2 - q^2)^{-1/2}$ by Eqs. (6.62) and (6.49)]. The extra term in Eq. (6.66) could be significant for a test particle with an appropriate charge-to-mass ratio, and in principle provides a new 5D test.

The last relation (6.66) and others we have derived in this section have concentrated on the classical effects of charge, insofar as we have taken a class of 5D charged black-hole solutions (6.42) with a classical test particle. The latter has $q/G^{1/2}m \ll 1$ with conventional units restored. Also of interest is the case $q/G^{1/2}m \gg 1$, which is typical of elementary particles and usually treated by quantum theory. That there is a problem in treating particles with large charge-to-mass ratios or fluids with large charge densities was appreciated by Kaluza [1]. However, a possible resolution is available through the observation that while conventional 4D causality is determined by $ds^2 \geq 0$, there is no rule in 5D which requires $dS^2 > 0$. It has therefore been suggested that in 5D the paths of particles have $dS^2 < 0$ [8] or $dS^2 = 0$ [18]. The latter proposition is especially attractive, because it agrees with the null nature of the 5D field equations, $R_{AB} = 0$. This possibility will be investigated in Chapter 7.

To finish the present section, we reconsider the metric (6.41) of a charged soliton, and ask what the energy of such an object would be if viewed from infinitely far away. This sounds simple, but is non-trivial to calculate even in 4D, let alone 5D. However, Hawking and others showed that in 4D the energy as defined from the action can be obtained also from the ADM formalism, wherein a metric is decomposed into separate parts called the lapse and shift, which allows of the formulation of energy terms which resemble those found in the Hamiltonian approach to classical mechanics. In 5D, it is possible to follow a similar procedure for the metrics which depends on the extra coordinate and are asymptotically de Sitter [19]. This is instructive for canonical metrics with a finite cosmological 'constant' [20], but can also be applied to the metric (6.41) for the charged soliton [21]. The expression for the energy involves integrals over the extra coordinate and a suitably-chosen compact 2-surface; but the working is tedious, so we give here only the result. It depends on the two parameters a, b which define the metric, and on the constant (6.45) defined above for the charge-to-mass ratio of the source, namely $k \equiv Q^2/4M^2(a - b)^2$ where $a \neq b$. The total energy is

$$E = M(a + b/2) - Mk(a/2 + b) \quad . \tag{6.67}$$

The first term here reproduces the results in the literature for the neutral soliton, while the second term is the modification due to the electric charge.

6.6. Field Equations and Induced Matter

The basis of Space-Time-Matter Theory is that the empty Ricci-flat field equations $R_{AB} = 0$ in 5D can be reduced to the Einstein equations with matter $G_{\alpha\beta} = 8\pi T_{\alpha\beta}$ in 4D. We discussed this in Chapter 2, and gave several examples where the induced matter of the effective energy-momentum tensor $T_{\alpha\beta}$ was calculated. However, these examples were for diagonal metrics, whereas electromagnetism typically involves the $g_{4\alpha}$ potentials so the metric is non-diagonal. In the present section, we wish to show how to do the calculation when the metric takes the standard electromagnetic form. We aim particularly at evaluating the electromagnetic part of T_{00} for the charged 5D black holes discussed in Sections 6.4 and 6.5 above.

The line element $dS^2 = g_{AB}dx^A dx^B$ for electromagnetic problems is usually taken as Eq. (6.10), which it is convenient to rewrite in matrix form:

$$g_{AB} = \begin{bmatrix} (g_{\alpha\beta} - \Phi^2 A_\alpha A_\beta) & -\Phi^2 A_\alpha \\ -\Phi^2 A_\beta & -\Phi^2 \end{bmatrix} ,$$

$$g^{AB} = \begin{bmatrix} g^{\alpha\beta} & -A^\alpha \\ -A^\beta & (-\Phi^{-2} + A^\mu A_\mu) \end{bmatrix} . \tag{6.68}$$

Here $A^\alpha = g^{\alpha\beta} A_\beta$ etc. as usual, and we use the same notation as elsewhere with hats added to the 4D parts of 5D quantities if there is a danger of confusion with the pure 4D quantities. For situations with electromagnetic fields, it is natural to consider the case where $g_{AB} = g_{AB}(x^\alpha)$. Then the nonzero 5D Christoffel symbols of the second kind are:

$$\hat{\Gamma}^\alpha_{\beta\gamma} = \Gamma^\alpha_{\beta\gamma} + \frac{1}{2}\Phi^2 \left(A_\beta F^\alpha{}_\gamma + A_\gamma F^\alpha{}_\beta \right) + \Phi\Phi^\alpha A_\beta A_\gamma ,$$

$$\Gamma^{\alpha}_{44} = \Phi\Phi^{\alpha} \ ,$$

$$\Gamma^{\alpha}_{\beta 4} = \frac{1}{2}\Phi^2 F^{\alpha}_{\ \beta} + \Phi\Phi^{\alpha}A_{\beta} \ ,$$

$$\Gamma^4_{44} = -\Phi\Phi^{\lambda}A_{\lambda} \ ,$$

(6.69)

$$\Gamma^4_{4\beta} = -\frac{1}{2}\Phi^2 A^{\lambda}F_{\lambda\beta} - \Phi\Phi^{\lambda}A_{\lambda}A_{\beta} + \Phi^{-1}\Phi_{\beta} \ ,$$

$$\Gamma^4_{\alpha\beta} = \frac{1}{2}\left(\nabla_{\beta}A_{\alpha} + \nabla_{\alpha}A_{\beta}\right) - \frac{1}{2}\Phi^2 A^{\lambda}\left(A_{\alpha}F_{\lambda\beta} + A_{\beta}F_{\lambda\alpha}\right)$$
$$- A_{\alpha}A_{\beta}\Phi\Phi^{\lambda}A_{\lambda} + \Phi^{-1}\left(A_{\alpha}\Phi_{\beta} + A_{\beta}\Phi_{\alpha}\right) \ .$$

Here $\Phi_{\alpha} \equiv \partial_{\alpha}\Phi$ and $\nabla_{\beta}A_{\alpha} = \partial_{\beta}A_{\alpha} - \Gamma^{\lambda}_{\alpha\beta}A_{\lambda}$ as usual. From Eqs. (6.69) we have $\Gamma^D_{\alpha D} = \Gamma^{\lambda}_{\alpha\lambda} + \Phi^{-1}\Phi_{\alpha}$ and $\Gamma^D_{4D} = 0$, where a repeated index indicates summation. Using these expressions with Eqs. (6.69), we can work out the components of the 5D Ricci tensor:

$$\hat{R}_{\alpha\beta} = R_{\alpha\beta} + \frac{1}{2}\Phi^2 F_{\alpha\lambda}F_{\beta}^{\ \lambda} - \Phi^{-1}\nabla_{\beta}\Phi_{\alpha}$$
$$+ \frac{1}{2}A_{\alpha}\left(\Phi^2\nabla_{\gamma}F^{\gamma}_{\ \beta} + 3\Phi\Phi^{\gamma}F_{\gamma\beta}\right)$$
$$+ \frac{1}{2}A_{\beta}\left(\Phi^2\nabla_{\gamma}F^{\gamma}_{\ \alpha} + 3\Phi\Phi^{\gamma}F_{\gamma\alpha}\right)$$
$$+ A_{\alpha}A_{\beta}\left(\frac{1}{4}\Phi^4 F^{\mu\nu}F_{\mu\nu} + \Phi\nabla_{\mu}\Phi^{\mu}\right) \ ,$$

(6.70)

$$R_{4\alpha} = \frac{1}{2}\Phi^2\nabla_{\lambda}F^{\lambda}_{\ \alpha} + \frac{3}{2}\Phi\Phi^{\lambda}F_{\lambda\alpha}$$
$$+ A_{\alpha}\left(\frac{1}{4}\Phi^4 F^{\mu\nu}F_{\mu\nu} + \Phi\nabla_{\mu}\Phi^{\mu}\right) \ ,$$

$$R_{44} = \frac{1}{4}\Phi^4 F^{\mu\nu}F_{\mu\nu} + \Phi\nabla_{\alpha}\Phi^{\alpha} \ .$$

These are written with the terms grouped in this fashion in order to simplify the field equations $R_{AB} = 0$, to which we will return below.

For now, we note that Eqs. (6.70) can be used to obtain the 4D induced or effective energy-momentum tensor. First we form the 5D quantities

$$\hat{R} \equiv g^{AB}R_{AB} = R + \frac{1}{4}\Phi^2 F^{\mu\nu}F_{\mu\nu} - 2\Phi^{-1}\nabla_{\mu}\Phi^{\mu} \ , \qquad (6.71)$$

$$\hat{R}^{\alpha\beta} \equiv \hat{g}^{\alpha A} \hat{g}^{\beta B} R_{AB}$$
$$= R^{\alpha\beta} + \frac{1}{2}\Phi^2 F^{\alpha}{}_{\lambda} F^{\beta\lambda} - \Phi^{-1}\nabla^{\beta}\Phi^{\alpha} \quad , \tag{6.72}$$

where $\nabla^{\beta} A^{\alpha} = g^{\beta\gamma}\nabla_{\gamma} A^{\alpha} = g^{\beta\gamma}(\partial_{\gamma} A^{\alpha} + \Gamma^{\alpha}_{\gamma\delta} A^{\delta})$ as usual. Then we find that the 4D part of the 5D Einstein tensor and the usual 4D Einstein tensor are related by:

$$\hat{G}^{\alpha\beta} \equiv \hat{R}^{\alpha\beta} - \frac{1}{2}\hat{g}^{\alpha\beta}\hat{R}$$
$$= R^{\alpha\beta} - \frac{1}{2}g^{\alpha\beta}R + \frac{1}{2}\Phi^2\left(F^{\alpha}{}_{\lambda}F^{\beta\lambda} - \frac{1}{4}g^{\alpha\beta}F^{\mu\nu}F_{\mu\nu}\right)$$
$$- \Phi^{-1}\left(\nabla^{\beta}\Phi^{\alpha} - g^{\alpha\beta}\nabla_{\mu}\Phi^{\mu}\right) \quad . \tag{6.73}$$

If we now impose the field equations $R_{AB} = 0$ or $G_{AB} = 0$, the first ten may be rewritten using Eq. (6.73) in the form of Einstein's equations:

$$R^{\alpha\beta} - \frac{1}{2}g^{\alpha\beta}R = 8\pi\left(T^{\alpha\beta}_{\text{em}} + T^{\alpha\beta}_{\text{scalar}}\right) \quad ,$$

$$T^{\alpha\beta}_{\text{em}} \equiv -\frac{1}{2}\Phi^2\left(F^{\alpha}{}_{\lambda}F^{\beta\lambda} - \frac{1}{4}g^{\alpha\beta}F^{\mu\nu}F_{\mu\nu}\right) \quad , \tag{6.74}$$

$$T^{\alpha\beta}_{\text{scalar}} \equiv \Phi^{-1}\left(\nabla^{\beta}\Phi^{\alpha} - g^{\alpha\beta}\nabla_{\mu}\Phi^{\mu}\right) \quad .$$

The last two quantities are the energy-momentum tensors as determined mainly by the electromagnetic and scalar fields, respectively.

Returning to the full set of fifteen field equations $R_{AB} = 0$, these may be rewritten using Eqs. (6.70) as three neat sets of ten, four and one equation each:

$$R_{\alpha\beta} = -\frac{1}{2}\Phi^2 F_{\alpha\lambda}F_{\beta}{}^{\lambda} + \Phi^{-1}\nabla_{\beta}\Phi_{\alpha} \quad ,$$

$$\nabla_{\lambda}F^{\lambda}{}_{\alpha} = -3\Phi^{-1}\Phi^{\lambda}F_{\lambda\alpha} \quad , \tag{6.75}$$

$$\nabla_{\alpha}\Phi^{\alpha} = -\frac{1}{4}\Phi^3 F^{\mu\nu}F_{\mu\nu} \quad .$$

In general, these may be regarded as Einstein-like equations for spacetime curved by electromagnetic and scalar fields, Maxwell-like equations where the source depends on the scalar field, and a wave-like equation for the scalar potential. They have wide applicability.

For present purposes, we look at Eqs. (6.75) for the case where the 4D part of the metric and the electromagnetic potential have the forms

$$ds^2 = e^\nu dt^2 - e^\lambda dr^2 - e^\mu r^2 (d\theta^2 + \sin^2\theta \, d\phi^2) \ ,$$

$$A_\alpha = (A_0, 0, 0, 0) \ . \tag{6.76}$$

Here the coefficients ν, λ, μ and A_0 depend only on r. In this case, the last member of Eqs. (6.75) reduces to

$$\Phi'' + \left(\frac{\nu' - \lambda' + 2\mu'}{2} + \frac{2}{r}\right)\Phi' - \frac{1}{2}\Phi^3 e^{-\nu}(A_0')^2 = 0 \ , \tag{6.77}$$

where a prime denotes the derivative with respect to r. Also the middle member of Eq. (6.75) can be once integrated to give

$$A_0' = -\frac{C}{r^2\Phi^3} e^{(\nu+\lambda-2\mu)/2} \ , \tag{6.78}$$

where C is an arbitrary constant. The first member of Eqs. (6.75) gives three relations:

$$\nu'' + \nu'\left(\frac{\nu'-\lambda'+2\mu'}{2} + \frac{2}{r}\right) = -\frac{\Phi'}{\Phi}\nu' - \Phi^2 e^{-\nu}(A_0')^2 \ ,$$

$$\nu'' + 2\mu'' + \frac{(\nu')^2 + 2(\mu')^2 - \lambda'(\nu'+2\mu')}{2} + \frac{4\mu'-2\lambda'}{r}$$
$$= -\Phi^2 e^{-\nu}(A_0')^2 - 2\Phi^{-1}\left(\Phi'' - \frac{\lambda'}{2}\Phi'\right) \ , \tag{6.79}$$

$$1 - \frac{1}{2}e^{\mu-\lambda}r^2\left[\mu'' + \frac{\mu'(\nu'-\lambda'+2\mu')}{2} + \frac{\nu'-\lambda'+4\mu'}{r} + \frac{2}{r^2}\right]$$
$$= \frac{1}{2}e^{\mu-\lambda}r^2\left(\mu' + \frac{2}{r}\right)\frac{\Phi'}{\Phi} \ .$$

The five equations (6.77)–(6.79) constitute a complete set for the five unknowns Φ, A_0, ν, λ and μ. It may be verified by hand that the charged soliton solutions (6.41) satisfy Eqs. (6.77)–(6.79), or by computer that they satisfy the equivalent set $R_{AB} = 0$.

The charged black-hole solutions (6.42) are an algebraically simple subset of the metric (6.41), and we are particularly interested in evaluating the energy density of the electrostatic field in this case. Putting Eqs. (6.42) into Eqs. (6.74) and using Eqs. (6.43)–(6.44) gives

$$
\begin{aligned}
T_{00}^{em} &= -\frac{1}{2}\Phi^2\left[F_{0\lambda}F_0^{\ \lambda} - \frac{1}{4}g_{00}F^{\mu\nu}F_{\mu\nu}\right] \\
&= -\frac{1}{2}E\left[g^{11}(F_{01})^2 - \frac{1}{4}g_{00}F^{\mu\nu}F_{\mu\nu}\right] \\
&= \frac{BM^2k}{E^3r^4} \quad .
\end{aligned}
\tag{6.80}
$$

By Eq. (6.45), $k = Q^2/4M^2$; and by Eqs. (6.46) we have $E \to 1$ and $B \to 1$ as $r \to \infty$. Thus, in this same limit $T_{00}^{em} \to Q^2/4r^4$. This result, quoted as Eq. (6.59) in Section 6.5, is of conventional form and serves as a consistency check of STM theory for electromagnetic fields.

6.7. The Charge and Mass of a Particle

In the present section we will investigate how far it is possible to explain the charge q and mass m of a particle in purely geometric terms. We believe this to be feasible, because the 5D geodesic was seen to give an expression for the energy of a charged particle (6.39) which closely resembles the 4D one (6.40). It turns out, however, that the best way to approach this problem is via the Lagrangian, and that in the last step of the analysis it is necessary to bring in the canonical metric of Chapter 7. We therefore approach the problem anew, following the route of Wesson and Liu [22]. We will occasionally come across relations which are familiar from the first part of this chapter, and the notation is the same. But we will not employ the field equations, proceeding by generalizing metric-based quantities from 4D to 5D.

In 4D, we have a line element which is given in terms of a metric tensor by $ds^2 = g_{\alpha\beta}dx^\alpha dx^\beta$, and a Lagrangian which is commonly defined as

$$L = m\frac{ds}{d\lambda} + qA_\mu \frac{dx^\mu}{d\lambda} \quad .$$ (6.81)

Here λ is a parameter and A_μ is the electromagnetic potential. This is acceptable, but clearly m and q are introduced *ad hoc*. Overlooking this for the moment, we form the momentum 4-vector

$$p_\alpha = \frac{\partial L}{\partial (dx^\alpha/d\lambda)} = mg_{\alpha\beta}\frac{dx^\beta}{ds} + qA_\alpha \quad .$$ (6.82)

That is, $p_\alpha = mg_{\alpha\beta}u^\beta + qA_\alpha$ and $p^\alpha \equiv g^{\alpha\beta}p_\beta = mu^\alpha + qA^\alpha$ where $u^\alpha \equiv dx^\alpha/ds$ is the 4-velocity. The product is

$$p_\alpha p^\alpha = m^2 u_\alpha u^\alpha + 2mqA_\alpha u^\alpha + q^2 A_\alpha A^\alpha \quad ,$$ (6.83)

which is just m^2 when we put $q = 0$ and choose the signature to be $g_{\alpha\beta} = (+1, -1, -1, -1)$. Locally, $g_{\alpha\beta} \approx \eta_{\alpha\beta}$, and for a charged particle moving with a 3-velocity $v^{123} = dx^{123}/dt$ we have

$$p^0 = m\frac{dt}{ds} + qA^0 = \frac{m}{(1-v^2)^{1/2}} + qA^0 \quad ,$$ (6.84)

$$p^{123} = m\frac{dx^{123}}{ds} + qA^{123} = \frac{mv^{123}}{(1-v^2)^{1/2}} + qA^{123} \quad ,$$ (6.85)

which we call the energy and (3D) momenta. These relations lead of course to the result $E^2 = p^2 + m^2$, which is the basis of particle physics and is in excellent agreement with experimental data. However, as pointed out above, there is no explanation of m or (if it is finite) of q.

In 5D, we have a line element which is given in terms of a metric tensor by $dS^2 = g_{AB}dx^A dx^B$. As in Eq. (6.10), we can rewrite this as the sum of a 4D part and an extra part that depends on scalar and vector fields:

$$dS^2 = g_{\alpha\beta}dx^\alpha dx^\beta - \Phi^2(d\ell + A_\alpha dx^\alpha)^2 \quad .$$ (6.86)

The corresponding Lagrangian is

$$L \equiv m\frac{dS}{d\lambda} = m\left[g_{\alpha\beta}\frac{dx^\alpha}{d\lambda}\frac{dx^\beta}{d\lambda} - \Phi^2\left(\frac{d\ell}{d\lambda} + A_\alpha\frac{dx^\alpha}{d\lambda}\right)^2\right]^{1/2} \quad .$$ (6.87)

Here m is still introduced *ad hoc,* but the theory has one extra parameter associated with the fifth coordinate ℓ and the charge q does not appear explicitly (see below). From Eq. (6.87) we can form the 4-part of the momentum 5-vector:

$$P_\alpha = \frac{\partial L}{\partial(dx^\alpha/d\lambda)} = m\left[g_{\alpha\beta}\frac{dx^\beta}{dS} + nA_\alpha\right] ,$$

$$n \equiv -\Phi^2\left(\frac{d\ell}{dS} + A_\alpha\frac{dx^\alpha}{dS}\right) . \qquad (6.88)$$

This is the 5D analog of the 4D relation (6.82). The extra or fifth component is

$$P_4 = mn . \qquad (6.89)$$

We see that the scalar function n of Eq. (6.88) specifies the momentum in the extra dimension. We might expect n to be small in many situations, because from Eqs. (6.86) and (6.88), we have $ds/dS = (1 + n^2/\Phi^2)^{1/2}$ [as in Eq. (6.22)], so $ds \approx dS$ in this case. To get the contravariant forms of the covariant vectors (6.88) and (6.89), we recall that

$$g^{AB} = \begin{bmatrix} g^{\alpha\beta} & -A^\alpha \\ -A^\alpha & (A^\mu A_\mu - \Phi^{-2}) \end{bmatrix} , \qquad (6.90)$$

with $A^\mu \equiv g^{\mu\nu}A_\nu$. This gives $P^\alpha \equiv g^{\alpha B}P_B = mU^\alpha$ and $P^4 \equiv g^{4B}P_B = mU^4$ where $U^A \equiv dx^A/dS$ is the 5-velocity. The product is

$$P_A P^A = m^2 . \qquad (6.91)$$

This is the 5D analog of the 4D relation (6.83). However, q does not appear in the former whereas it does in the latter, so we need to inquire where this parameter fits into the extended geometry. To answer this question, we rewrite Eq. (6.88) as

$$P_\alpha = m\left[g_{\alpha\beta}U^\beta + nA_\alpha\right] . \qquad (6.92)$$

Although it is the 4-part of a 5-vector, the metric (6.86) from which it is derived is invariant under the 4D subset of general 5D coordinate transformations $x^\alpha \to \bar{x}^\alpha(x^\beta)$. Eq. (6.92) must inherit the same property. That is, P_α is a 4-vector. To recast it in 4D form, we again apply $ds/dS = (1 + n^2/\Phi^2)^{1/2}$ from above, and find that

$$P_\alpha = \left[m \left(g_{\alpha\beta} u^\beta + \frac{n}{(1+n^2\Phi^{-2})^{1/2}} A_\alpha \right) \right] \frac{ds}{dS} \ . \tag{6.93}$$

This can be compared with the purely 4D relation (6.82):

$$p_\alpha = m \left(g_{\alpha\beta} u^\beta + \frac{q}{m} A_\alpha \right) \ . \tag{6.94}$$

Clearly the vectors are related via $P_\alpha = p_\alpha(ds/dS)$, so the charge-to-mass ratio of the test particle is

$$\frac{q}{m} = \frac{n}{(1+n^2\Phi^{-2})^{1/2}} \ . \tag{6.95}$$

This means that charge is equal to mass times a function of coordinates, so at this point we are halfway to explaining both parameters in terms of geometry.

The result (6.95) derived here from the Lagrangian agrees with Eq. (6.29), derived before from the 4D part of the 5D geodesic. The last was studied extensively above, and to speed the present analysis we recall that the motion in the $x^4 = \ell$ and $x^0 = t$ directions can be summed up in terms of two pairs of relations:

$$\frac{dn}{dS} = \frac{1}{2} \frac{\partial g_{AB}}{\partial \ell} \frac{dx^A}{dS} \frac{dx^B}{dS} \ ,$$

$$n \equiv -\Phi^2 \left(\frac{d\ell}{dS} + A_\alpha \frac{dx^\alpha}{dS} \right) \ , \tag{6.96}$$

$$\frac{de}{dS} = \frac{1}{2} \frac{\partial g_{AB}}{\partial t} \frac{dx^A}{dS} \frac{dx^B}{dS} \ ,$$

$$e \equiv \frac{g_{00}^{1/2}}{(1-v^2)^{1/2}} \left(1 + \frac{n^2}{\Phi^2} \right)^{1/2} + nA_0 \ . \tag{6.97}$$

The functions n, e are constants of the motion if the metric is independent of ℓ, t respectively; and as noted before, the second of Eqs. (6.97) is strongly reminiscent of the expression for the energy of a test particle in a static 4D spacetime:

$$\mathsf{E} = \frac{g_{00}^{1/2}}{(1-v^2)^{1/2}} m + q A_0 \ . \tag{6.98}$$

The route to explaining charge and mass as geometrical objects is now apparent, insofar as we expect the q, m of 4D to be related to the n, e of 5D, subject to compatibility with Eq. (6.95) for the ratio.

Let us look at the mass first, for the case of an uncharged test particle $(q, n \to 0)$. In Eq. (6.97), g_{00} refers to a 5D metric coefficient, and we take the canonical 5D metric in the weak-field case where the 4D part is close to flat, so $dS^2 \approx (\ell/L)^2 ds^2 - d\ell^2$ with $ds^2 = \eta_{\alpha\beta} dx^\alpha dx^\beta$ and $\eta_{\alpha\beta} = (+1, -1, -1, -1)$. Then $\sqrt{g_{00}} = \ell/L$ in Eq. (6.97). In (6.98), g_{00} refers to a 4D metric coefficient, and in the corresponding limit $\sqrt{g_{00}} = 1$. Then Eqs. (6.97) and (6.98) read respectively

$$e = \frac{\ell}{L(1-v^2)^{1/2}} \quad , \quad \mathsf{E} = \frac{m}{(1-v^2)^{1/2}} \ . \tag{6.99}$$

Here e is dimensionless, while E has the dimensions of a length (due to the *ad hoc* introduction of m in 4D). Thus we make the identification

$$\mathsf{E} \equiv eL = \frac{\ell}{(1-v^2)^{1/2}} \ . \tag{6.100}$$

This means that the role of the 4D uncharged rest mass is played in 5D geometry by the extra coordinate.

Now let us look at the charge. A comparison of Eqs. (6.96), (6.97) and (6.98) shows that we have to make a distinction between the mass of an uncharged particle (which we have above labeled m) and the mass of a charged particle (which we here label m_q). The appropriate definitions are

$$m_q \equiv \ell \left(1 + \frac{n^2}{\Phi^2} \right)^{1/2} \quad , \quad q \equiv n\ell \ . \tag{6.101}$$

These mean that the mass is augmented by the electromagnetic self-energy of the particle, as one might expect; but more importantly that the role of 4D charge is played in 5D geometry by the extra momentum, as originally proposed by Klein.

The ratio q/m_q derived from Eq. (6.101) agrees with Eq. (6.95), and in fact a more detailed investigation of the relations we have discussed in this section shows there is little room for algebraic maneuvers [22]. In canonical coordinates, the rest mass of an uncharged particle is ℓ, the mass of a charged particle is $\ell\sqrt{1 + n^2/\Phi^2}$, and the charge itself is $n\ell$. The charge-to-mass ratio which appears in the Lorentz force law is $q/m_q = n(1 + n^2/\Phi^2)^{-1/2}$. All of these 4D physical quantities can be derived from 5D geometrical ones.

6.8. Conclusion

The 5D metric with gravitational, electromagnetic and scalar potentials is commonly written in the form of Eq. (6.10) to distinguish these fields. The geodesic equation gives a new scalar associated with motion in the extra dimension (6.21), which comparison with the motion in spacetime (6.28) reveals to be connected with the charge-to-mass ratio of a test particle, Eq. (6.29). However, the new scalar is not a constant of the motion unless the metric is independent of x^4, and the charge-to-mass ratio is in general a function of all five coordinates. The time component of the 5D geodesic defines a scalar (6.39) analogous to the particle energy (6.40) in general relativity.

The static, neutral solitons which we met before as the standard 1-body solutions in 5D can be generalized to a 3-parameter class with charge given by Eq. (6.41). In the case where the embedded 4-space in the neutral limit is Schwarzschild, we can speak in the general case of a 2-parameter class of charged 5D black holes defined by Eqs. (6.42). The motion of a test particle in orbit around such an object shows interesting effects, but even when the 5D metric becomes effectively 4D, as in Eq. (6.57), it does not exactly reproduce the Reissner-Nordström metric (6.58) of general relativity. The field equations for x^4-independent electromagnetic problems are given in Eqs. (6.75) and have wide applicability, though the energy density of the electrostatic field for the charged black holes has the standard form, Eq. (6.80). Eq. (6.101) shows that it is possible to obtain explicit expressions for the charge and mass of a particle, a result which may come as a surprise to workers in classical electromagnetism.

Electromagnetic effects in 5D have been somewhat neglected compared to those in other areas, such as cosmology. The few studies which have been made concern mainly particle effects. Notably, modifications to the properties of the vacuum and the scattering matrix [23, 24]. It is well-known that there are similarities between certain solutions of general relativity and the properties of some elementary particles [25, 26]. However, the range of potential solutions which might be matched to particles is far larger in 5D than in 4D, and accordingly deserves attention.

References

[1] Kaluza, T., *Sitz. Preuss. Akad. Wiss.* **33**, 966 (1921).

[2] Klein, O., *Z. Phys.* **37**, 895 (1926).

[3] Wesson, P.S. and Ponce de Leon, J., *Astron. Astrophys.* **294**, 1 (1995).

[4] Landau, L.D. and Lifshitz, E.M., *The Classical Theory of Fields* (Pergamon, Oxford, 1975), 251.

[5] Leibowitz, E. and Rosen, N., *Gen. Rel. Grav.* **4**, 449 (1973).

[6] Kovacs, D., *Gen. Rel. Grav.* **16**, 645 (1984).

[7] Gegenberg, J. and Kunstatter, G., *Phys. Lett.* **A106**, 410 (1984).

[8] Davidson, A. and Owen, D.A., *Phys. Lett.* **B177**, 77 (1986).

[9] Ferrari, J.A., *Gen. Rel. Grav.* **21**, 683 (1989).

[10] Mashhoon, B., Liu, H. and Wesson, P.S., *Phys. Lett.* **B331**, 305 (1994).

[11] Mashhoon, B., Wesson, P.S. and Liu, H., *Gen. Rel. Grav.* **30**, 555 (1998).

[12] Appelquist, T. and Chodos, A., *Phys. Rev.* **D28**, 772 (1993).

[13] Overduin, J.M. and Wesson, P.S., *Phys. Rep.* **283**, 303 (1997).

[14] Liu, H. and Wesson, P.S., *Phys. Lett.* **B381**, 420 (1996).

[15] Liu, H. and Wesson, P.S., *Class. Quant. Grav.* **14**, 1651 (1997).

[16] Chatterjee, S., *Astron. Astrophys.* **230**, 1 (1990).

[17] Kalligas, D., Wesson, P.S. and Everitt, C.W.F., *Astrophys. J.* **439**, 548 (1995).

[18] Seahra, S.S. and Wesson, P.S., *Gen. Rel. Grav.* **33**, 1731 (2001);
     Youm, D., *Mod. Phys. Lett.* **A16**, 2371 (2001).

[19] Sajko, W.N., *Phys. Rev.* **D60**, 104038 (1999).

[20] Sajko, W.N. and Wesson, P.S., *Gen. Rel. Grav.* **32**, 1381 (2000).

[21] Sajko, W.N. and Wesson, P.S., *Mod. Phys. Lett.* **A16**, 627 (2001).

[22] Wesson, P.S. and Liu, H., *Int. J. Theor. Phys.* **36**, 1865 (1997).

[23] Nodvik, J.S., *Phys. Rev. Lett.* **55**, 2519 (1985).

[24] Liko, T., *Phys. Lett.* **B617**, 193 (2005).

[25] Arcos, H.J., and Pereira, J.G., in Petkov, V. (ed.), *Relativity and the Dimensionality of the World* (Springer, Dordrecht, 2007).

[26] 't Hooft, G., *Found. Phys.* **41**, 1829 (2011).

Chapter 7

Canonical Space

7.1. Introduction

The universe appears to be dominated by the vacuum, and canonical space is the 5D metric whose 4D part gives back the classical vacuum of general relativity. In that theory, the effective properties of the vacuum are measured by the cosmological constant, which will be the focus of the next chapter. In the present one, we will concentrate on the other attributes of canonical space.

It is a fact that 5D canonical space contains all vacuum solutions of Einstein's equations (meaning those empty of ordinary matter but with an energy density and pressure proportional to the cosmological constant). This follows from Campbell's theorem (see Section 2.5). Canonical space was introduced by Mashhoon, Liu and Wesson in 1994 [1]. They modelled the metric on the synchronous one in 4D standard cosmology. In some ways it is the analog in STM theory of the warp metric in M theory, though the two are not identical. The appellation 'canonical' was suggested by the simple algebraic form of the metric, and by the fact that it led to drastic simplification in the 5D field equations and their corresponding equations of motion (see below). In crude terms, the simplification afforded by the 5D canonical metric is akin to the way in which a 3D metric in Cartesian coordinates xyz becomes more transparent for certain situations when expressed in polar coordinates $r\theta\phi$ (this is why problems in astrophysics and geophysics are usually treated in terms of the latter coordinates rather than the former). The general form of 5D canonical space consists of the square of the extra coordinate (divided by a length to preserve physical dimensions) multiplied onto the usual 4D interval, plus an extra flat piece. This 5D form is algebraically general, provided the 4D interval (metric tensor) is allowed to depend on the extra coordinate. It is arrived at by using the five coordinate degrees of freedom to eliminate the off-diagonal (electromagnetic) potentials, and to set the magnitude of the extra diagonal potential to unity. This last condition is rather restrictive, and similar to the one used by Kaluza and Klein in the early years of 5D physics (see Section 2.2). However, it is acceptable, provided it is realized that thereby all of the new physics associated with the fifth dimension is concentrated into the extra coordinate and its presence as a quadratic factor multiplied onto spacetime. It is therefore necessary to keep an open mind about the meaning of the extra coordinate when working out problems in 5D using the canonical metric. However, x^4 has a simple interpretation when the 4D part of the metric (after the quadratic factor) is independent of it, the so-called pure-canonical case. The dynamics then shows that x^4 is equal (up to a constant) with the rest mass of a test particle moving in the field described by the metric. This is in agreement with one interpretation of Mach's principle (see Section 8.2), according to which the mass of a particle measured locally ought to be dependent on its surroundings.

In treating the properties of canonical space in the following sections, the general form of the metric will be denoted C_5, while the special case where the (post-factored) spacetime is independent of the extra coordinate will be denoted C_5^*.

7.2. The Metric and the Cosmological Constant

We established in Chapter 2 that the 5D Ricci-flat field equations $R_{AB} = 0$ contain the 4D Einstein equations $G_{\alpha\beta} = 8\pi T_{\alpha\beta}$. The exact form of the induced or effective energy-momenta tensor was found to be given by Eq. (2.26). That expression contains in general contributions from all types of matter, including the vacuum. In Einstein's theory, the latter behaves as a perfect fluid with an equation of state $p = -\rho = -\Lambda/8\pi$, where Λ is the cosmological constant. Alternatively, if there is no ordinary matter present, the Einstein equations read $G_{\alpha\beta} = -\Lambda g_{\alpha\beta}$. By using these equations, or the vacuum limit of the energy-momenta tensor, it is possible to identify the 4D cosmological constant in the 5D theory.

The canonical metric and its associated cosmological constant are by now well known. We write the metric in its general form (C_5), and the cosmological constant for its special or pure form (C_5^*), thus:

$$dS^2 = \frac{\ell^2}{L^2} g_{\alpha\beta}(x^\gamma, \ell) dx^\alpha dx^\beta + \varepsilon d\ell^2 \ ,$$

$$\Lambda = \frac{-3\varepsilon}{L^2} \ , \quad g_{\alpha\beta} = g_{\alpha\beta}(x^\gamma \text{ only}) \quad .$$

(7.1)

Here L is a constant length. It serves to make the metric consistent in terms of its physical dimensions (units), and to fix the magnitude of the cosmological constant. This is positive for a spacelike extra dimension ($\varepsilon = -1$) and negative for a timelike extra dimension. The 4D line interval or proper time is $ds^2 = g_{\alpha\beta}(x^\gamma, \ell) dx^\alpha dx^\beta$, and looks like the conventional expression on the hypersurface $\ell = L$.

In view of the significance of Eq. (7.1) for vacuum physics, it is important to ask about its uniqueness, especially for the special form. This can be elucidated by using the field equations $R_{AB} = 0$, notably the component $R_{44} = 0$ (the precise form of this may be found in various places, including Section 4.2). We consider a metric like Eq. (7.1) but with a generic form:

$$dS^2 = f(\ell)ds^2 \pm g(\ell)d\ell^2 \quad .$$

(7.2)

The question then is: what constraint arises from $R_{44} = 0$ between the embedding function $f(\ell)$ and the scalar field $g(\ell)$? The answer, after some algebra, is

$$g = \left(\frac{\partial f}{\partial \ell}\right)^2 \frac{L^2}{f} \quad .$$

(7.3)

Here L is the arbitrary constant of integration which enters from the first integral of the second-order field equation, and is essentially the same Λ-defining parameter as in Eq. (7.1). The constraint imposed by Eq. (7.3) makes it straightforward to absorb the scalar field in Eq. (7.2) by a coordinate transformation, thereby fixing the form of the embedding function. The new x^4 coordinate is defined up to a second arbitrary constant of integration, such that $f \propto (\ell - \ell_0)^2$. Setting the second constant to zero and redefining the first one gives

$$dS^2 = \left(\frac{\ell}{L}\right)^2 g_{\alpha\beta}(x^\gamma) dx^\alpha dx^\beta \pm d\ell^2 \quad . \tag{7.4}$$

This is the (pure) canonical metric as often used in the literature, and is seen to be unique up to coordinate transformations in $x^4 = \ell$.

Further insight into 5D canonical space comes from comparing it to 4D de Sitter space. The latter can be regarded as a 4D pseudosphere of constant curvature, $-1/L^2$ for $\Lambda > 0$ and $+1/L^2$ for $\Lambda < 0$ [2]. Alternatively, the 4D metric

$$ds^2 = (1 - \Lambda r^2/3)dt^2 - (1 - \Lambda r^2/3)^{-1}dr^2 - r^2 d\Omega^2 \tag{7.5}$$

can be mapped to the 5D metric

$$dS^2 = dT^2 - (dX^2 + dY^2 + dZ^2) \pm dL^2 \quad . \tag{7.6}$$

Here the signature is $(+ - - - -)$ for $\Lambda > 0$ and $(+ - - - +)$ for $\Lambda < 0$. Thus, 4D de Sitter space is isometric to 5D Minkowski space. The physical cosmological constant Λ measures the radius of curvature L which necessarily appears in the algebraic passage from flat 5D to curved 4D. For $\Lambda < 0$, the mapping of de Sitter space to a pseudosphere means that a line around the centre can repeatedly traverse the circumference $2\pi L$. This is sometimes seen as a violation of 4D causality since events repeat with a period $2\pi L/c$. If desired, the behaviour can be avoided by supposing that events occur on different surfaces of a tightly-packed 'scroll'. However, from the 5D perspective this behaviour is natural, and can be usefully applied to particle physics, as we will see in Chapter 8.

In gravitation and particle physics, we noted in Chapter 1 that the rest mass m of a test object can be geometrized in two complementary ways. One uses the gravitational constant G and the Schwarzschild radius, while the other uses Planck's constant h and the Compton wavelength.

These define what may be called Einstein and Planck measures for the mass: $\ell_E \equiv Gm/c^2$, $\ell_P \equiv h/mc$. Since these are both lengths, they are convenient ways to handle the mass, even if they have no more profound meanings. However, in introducing canonical coordinates, Mashhoon *et al.* had in mind the identification of x^4 with the preceding gravitational measures of the mass [1]. The general definition of particle mass in 5D relativity is a subtle problem, as mentioned elsewhere. Here we confine ourselves to pointing out that the Einstein and Planck measures of mass are connected by a simple relation which may be viewed as a duality or a coordinate transformation: $\ell \to L^2/\ell$ where L is a constant length like the one introduced above. In fact, carrying out this transformation on the pure-canonical metric (7.4) gives

$$dS^2 = \left(\frac{L}{\ell}\right)^2 g_{\alpha\beta}(x^\gamma)dx^\alpha dx^\beta \pm \left(\frac{L}{\ell}\right)^4 d\ell^2 \ . \tag{7.7}$$

This form, compared to the original, basically inverts the prefactor and adds an ℓ-dependent scalar field. The latter, of course, being derived from a coordinate transformation, will not 'upset' the field equations.

A more significant coordinate transformation is the straight shift, $\ell \to (\ell - \ell_0)$ where ℓ_0 is a constant. This looks innocuous, but turns out to have a drastic effect on the cosmological 'constant' Λ. This might be suspected, because we have noted elsewhere that 4D physics is not generally covariant in the conventional way under a coordinate transformation that involves $x^4 = \ell$. To investigate this, Mashhoon and Wesson carried out two similar but lengthy analyses of the 5D field equations with a shift along the ℓ-axis [3]. The result is that what may be termed the shifted (pure) canonical metric has

$$dS^2 = \left(\frac{\ell-\ell_0}{L}\right)^2 g_{\alpha\beta}(x^\gamma)dx^\alpha dx^\beta + \varepsilon d\ell^2 \ ,$$

$$\Lambda = \frac{-3\varepsilon}{L^2}\left(\frac{\ell}{\ell-\ell_0}\right)^2 \ . \tag{7.8}$$

This can be compared to the unshifted form in Eq. (7.1): a divergence has appeared at $\ell = \ell_0$ in a cosmological 'constant' which is now ℓ-

dependent. Since the path of a particle in the metric (7.8) will in general be a function of the proper time s (see below), and since s is a function of the spacetime coordinates, we see that $\Lambda = \Lambda(x^\gamma)$. That is, Λ is no longer a constant but a variable or dynamical quantity. The result (7.8) has been confirmed using a different method by Ponce de Leon [4]. It arises fundamentally from the behaviour of the 4D Ricci tensor and Ricci scalar under the shift in ℓ. The shift $\ell \rightarrow (\ell - \ell_0)$ acts like a constant conformal transformation from the 4D perspective; and while R_{AB} is invariant under such a change, the quantity R is not because it is contracted with the (new) metric tensor [5]. Put another way, the Einstein equations in vacuum $R_{\alpha\beta} = \Lambda g_{\alpha\beta}$ have to hold both before and after the shift, with the unshifted and shifted values of Λ. Eq. (7.8) has major implications for cosmology and particle physics, which we mentioned in Chapter 3 and will take up in detail in Chapter 8.

7.3. Equations of Motion

The simplest way to obtain dynamics in 5D is to assume that all particles move on null paths with $dS^2 = 0$. It may be shown that this includes not only the paths of photons with $ds^2 = 0$, but also the paths of massive particles on 4D timelike paths with $ds^2 > 0$ [6, 7]. The assumption of a null 5D interval matches the null nature of the field equations $R_{AB} = 0$, and agrees with the inference from quantum field theory that *all* particles are essentially photon-like, with their masses related dynamically by a mechanism such as the Higgs field. But while the null-path assumption is simple, it only gives explicit results when the metric has a correspondingly simple form; and though we will use it later for the pure-canonical case, the general canonical metric (7.1) requires a more thorough investigation.

We follow the working of Mashhoon, Wesson and Liu [8]. These authors, it should be noted, use a normalization condition for the 4-velocities $u^\alpha \equiv dx^\alpha/ds$ which is formally the same as the one in general relativity, namely $u^\alpha u_\alpha = 1$. This is acceptable, but not unique, and other workers approach the subject in other ways. Also, the just-noted normalization condition leads, after covariant differentiation, to the well-known orthogonality relation $u^\mu f_\mu = 0$. Here f_μ is commonly called a force, but it is actually a force per unit mass or acceleration. The distinction makes no difference if a test particle has constant mass, as usually assumed in general relativity; but it does if the mass can be variable, as in 5D theory. The analog of the orthogonality relation in 5D is $u^A f_A = 0$, which implies $u^\mu f_\mu = -u^4 f_4 \neq 0$. We expect, therefore, to find an extra or fifth force in 5D dynamics. This is so in both STM theory and M theory [9, 10]. However, we will see below that this disappears in one particular coordinate frame, namely that for pure canonical space.

The equations of motion are obtained by minimizing the distance between two points in 5D, via $\delta[\int dS] = 0$. If λ is an arbitrary affine parameter along the path, this relation can be written $\delta[\int \mathcal{L} d\lambda] = 0$, where using Eq. (7.1) the quantity which is commonly referred to as the Lagrangian is

$$\mathcal{L} \equiv \frac{dS}{d\lambda} = \left[\frac{\ell^2}{L^2} g_{\alpha\beta} \frac{dx^\alpha}{d\lambda} \frac{dx^\beta}{d\lambda} - \left(\frac{d\ell}{d\lambda}\right)^2\right]^{1/2} . \tag{7.9}$$

In this the extra dimension is taken to be spacelike, though the opposite case is easy to obtain by appropriate sign changes in the results we will obtain. The path in 5D is described by $x^\alpha = x^\alpha(\lambda), \ell = \ell(\lambda)$. The momenta and the equations of motion are then given as usual by

$$p_\alpha = \frac{\partial \mathcal{L}}{\partial(dx^\alpha/d\lambda)} \quad , \quad p_\ell = \frac{\partial \mathcal{L}}{\partial(d\ell/d\lambda)} \quad , \tag{7.10}$$

$$\frac{dp_\alpha}{d\lambda} = \frac{\partial \mathcal{L}}{\partial x^\alpha} \quad , \quad \frac{dp_l}{d\lambda} = \frac{\partial \mathcal{L}}{\partial \ell} . \tag{7.11}$$

Defining $u^\alpha \equiv dx^\alpha/d\lambda, u^\ell \equiv d\ell/d\lambda$ and $\theta \equiv \mathcal{L}^2 = (\ell^2/L^2)g_{\alpha\beta}u^\alpha u^\beta - (u^\ell)^2$, we then have for the momenta

$$p_\alpha = \left(\frac{\ell^2}{L^2}\right)\frac{g_{\alpha\beta}u^\beta}{\sqrt{\theta}} \quad , \quad p_\ell = -\frac{u^\ell}{\sqrt{\theta}} . \tag{7.12}$$

The corresponding equations of motion are

$$\frac{d}{d\lambda}\left(g_{\alpha\beta}u^\beta\right) + \frac{2}{\ell}\left(\frac{d\ell}{d\lambda}\right)g_{\alpha\beta}u^\beta - g_{\alpha\beta}u^\beta\left(\frac{1}{2\theta}\right)\frac{d\theta}{d\lambda}$$

$$= \frac{1}{2}\frac{\partial g_{\beta\gamma}}{\partial x^\alpha}u^\beta u^\gamma \quad , \tag{7.13}$$

$$\frac{d^2\ell}{d\lambda^2} - \left(\frac{1}{2\theta}\right)\frac{d\theta}{d\lambda}\frac{d\ell}{d\lambda} = -\frac{\ell}{L^2}g_{\alpha\beta}u^\alpha u^\beta - \frac{1}{2}\left(\frac{\ell^2}{L^2}\right)\frac{\partial g_{\beta\gamma}}{\partial \ell}u^\beta u^\gamma \quad . \tag{7.14}$$

We interject at this point the observation that the (conserved) super-Hamiltonian is

$$H \equiv p_\alpha \frac{dx^\alpha}{d\lambda} + p_\ell \frac{d\ell}{d\lambda} - \mathcal{L} = 0 \quad , \tag{7.15}$$

as expected. It is now convenient to choose $\lambda = s$, so $g_{\alpha\beta}u^\alpha u^\beta = 1$. This normalization condition, it is important to note, applies to the metric after it has been factored by the typical canonical quadratic. Proceeding, we can now manipulate the equations of motion into the form

$$\frac{du^\mu}{ds} + \Gamma^\mu_{\beta\gamma}u^\beta u^\gamma = -g^{\mu\alpha}\frac{\partial g_{\alpha\beta}}{\partial \ell}\frac{d\ell}{ds}u^\beta + \left[\frac{1}{2\theta}\frac{d\theta}{ds} - \frac{2}{\ell}\frac{d\ell}{ds}\right]u^\mu \quad , \tag{7.16}$$

$$\frac{d^2\ell}{ds^2} - \left(\frac{1}{2\theta}\right)\frac{d\theta}{ds}\frac{d\ell}{ds} = -\frac{\ell}{L^2} - \frac{1}{2}\left(\frac{\ell^2}{L^2}\right)\frac{\partial g_{\beta\gamma}}{\partial \ell}u^\beta u^\gamma \quad . \tag{7.17}$$

Here $\Gamma^\mu_{\beta\gamma}$ is the usual 4D Christoffel symbol of the second kind. With the parameter along the path as 4D proper time, we have $\theta = (\ell^2/L^2) - (d\ell/ds)^2$, and the last equation can be written as

$$\frac{d^2\ell}{ds^2} - \frac{1}{\ell}\left(\frac{d\ell}{ds}\right)^2 = -\left[\frac{\ell^2}{L^2} - \left(\frac{d\ell}{ds}\right)^2\right]\left(\frac{1}{\ell} + \frac{1}{2}\frac{\partial g_{\beta\gamma}}{\partial \ell}u^\beta u^\gamma\right) \quad . \tag{7.18}$$

Substituting for θ in Eq. (7.16) and using Eq. (7.18), the 4D equations of motion can be written

$$\frac{du^\mu}{ds} + \Gamma^\mu_{\beta\gamma}u^\beta u^\gamma = F^\mu \quad , $$

$$\tag{7.19}$$

$$F^\mu \equiv \left(-g^{\mu\alpha} + \frac{1}{2}\frac{dx^\mu}{ds}\frac{dx^\alpha}{ds}\right)\frac{d\ell}{ds}\frac{dx^\beta}{ds}\frac{\partial g_{\alpha\beta}}{\partial \ell} \quad . $$

We see from Eqs. (7.19) that the usual geodesic of general relativity is modified by an extra or fifth force.

The extra component of the equation of motion (7.18) can be rewritten as

$$\frac{d^2\ell}{ds^2} - \frac{2}{\ell}\left(\frac{d\ell}{ds}\right)^2 + \frac{\ell}{L^2} = -\frac{1}{2}\left[\frac{\ell^2}{L^2} - \left(\frac{d\ell}{ds}\right)^2\right]\frac{dx^\alpha}{ds}\frac{dx^\beta}{ds}\frac{\partial g_{\alpha\beta}}{\partial\ell} \quad . \tag{7.20}$$

This equation figures in many physical applications. It can be once integrated in terms of a quantity $Z \equiv u^\alpha u^\beta (\partial g_{\alpha\beta}/\partial\ell)$, giving

$$\left(\frac{d\ell}{ds}\right)^2 = \ell^2 - \frac{\ell^4}{\ell_0^2}\exp\left(\int_{\ell_0}^{\ell} Z\,d\ell\right) \quad . \tag{7.21}$$

This gives the square of the velocity in the extra dimension in terms of the 4D interval and a fiducial value of the extra coordinate ($d\ell/ds = 0$ at $\ell = \ell_0$, where this usage is not to be confused with that in Section 7.2). Solutions to Eqs. (7.20) and (7.21) depend in general on how the post-factored metric depends on the extra coordinate and on the 4-velocities in spacetime. However, Eq. (7.20) is also satisfied when both scales go to zero, and $\ell = \ell_* \exp(\pm s/L)$, where ℓ_* is a constant. This is actually the solution for the null path $dS^2 = 0$ in the (pure) canonical metric, and is a kind of generic solution.

The new force (per unit mass) in the 4D equations of motion (7.19) is necessarily connected to the existence of the fifth dimension. For from Eq. (7.19) we can form the scalar quantity

$$F^\mu u_\mu = -\frac{1}{2}u^\alpha u^\beta \frac{\partial g_{\alpha\beta}}{\partial\ell}\frac{d\ell}{ds} \quad . \tag{7.22}$$

This has no 4D analog. Another way to see what is involved here is to start from the 4D scalar $u^\alpha u_\alpha = 1$, where we recall that in our approach this means $u^\alpha(s)g_{\alpha\beta}(x^\alpha,\ell)u^\beta(s) = 1$. The application of the total derivative D/ds to this, along with the observation that $Dg_{\alpha\beta}/ds = (\partial g_{\alpha\beta}/\partial\ell)(d\ell/ds)$, results in

$$F^\alpha g_{\alpha\beta}u^\beta + u^\alpha u^\beta \frac{\partial g_{\alpha\beta}}{\partial\ell}\frac{d\ell}{ds} + u^\alpha g_{\alpha\beta}F^\beta = 0 \quad . \tag{7.23}$$

This rearranged gives back Eq. (7.22). We see that the existence of the fifth force F^μ, and its associated scalar $F^\mu u_\mu$ (which is a kind of power), follows from the fact that we have used the 4D interval s to characterize the motion in a 5D manifold. This makes sense from a practical point of view, because we have a body of 4D physics which we wish to interpret in terms of 5D geometry. However, it is important to realize that $F^\mu u_\mu \neq 0$ is, in principle, a unique indication of the existence of at least one higher dimension.

To appreciate this, consider for the moment 4D Einstein gravity with other known forces. For example, the Lorentz force of electrodynamics obeys the relation $Du^\mu/ds = (q/m)F^\mu{}_\nu u^\nu$ where q/m is the charge-to-mass ratio of a test particle and $F^\mu{}_\nu$ is the Faraday field tensor. The antisymmetry of the latter means that $F_{\mu\nu}u^\mu u^\nu = 0$, so $F_\mu = (q/m)F_{\mu\nu}u^\nu$ obeys $F_\mu u^\mu = 0$. As another example, the force due to the pressure of a perfect fluid can be seen to obey a similar relation by considering the energy-momentum tensor $T^{\mu\nu} = (\rho + p)u^\mu u^\nu - pg^{\mu\nu}$. This obeys $\nabla_\nu T^{\mu\nu} = 0$ or

$$\partial_\nu(\rho + p)u^\mu u^\nu + (\rho + p)\nabla_\nu u^\mu u^\nu + (\rho + p)u^\mu \nabla_\nu u^\nu$$
$$- \partial_\nu p\, g^{\mu\nu} = 0 \ . \tag{7.24}$$

This can be multiplied by u_μ, and the term $(\rho + p)\nabla_\nu u^\mu u^\nu u_\mu$ dropped, because by taking the covariant derivative of $u^\mu u_\mu = 1$, we have $\nabla_\nu u^\mu u_\mu = 0$ (see above). Then there results

$$(\rho + p)\nabla_\nu u^\nu = -\partial_\nu p\, u^\nu \ , \tag{7.25}$$

which put back into Eq. (7.24) gives

$$F^\mu \equiv \nabla_\nu u^\mu u^\nu = \frac{\partial_\nu p\, (g^{\mu\nu} - u^\mu u^\nu)}{(\rho + p)} \ . \tag{7.26}$$

Again by virtue of $u^\mu u_\mu = 1$, this implies $F^\mu u_\mu = 0$. This result and the preceding one are not so much meant as a reminder of known physics, as illustrations of the fact that timelike motions in 4D general relativity with other known forces obey equations of motion of the form $Du^\mu/ds = F^\mu$ with $F_\mu u^\mu = 0$.

Conversely, the new force defined by Eq. (7.19) in this section has $F_\mu u^\mu \neq 0$. Thus at least part of the new force (per unit inertial mass) F^μ must be non-4D in origin. To see this explicitly, we can express F^μ as the sum of a component (N^μ) normal to the 4-velocity u^μ of the particle and a component (P^μ) parallel to it. Thus $F^\mu = N^\mu + P^\mu$, where

$$N^\mu = (-g^{\mu\alpha} + u^\mu u^\alpha) u^\beta \frac{\partial g_{\alpha\beta}}{\partial \ell} \frac{d\ell}{ds} \quad , \tag{7.27}$$

$$P^\mu = -\frac{1}{2} u^\mu \left(u^\alpha u^\beta \frac{\partial g_{\alpha\beta}}{\partial \ell} \right) \frac{d\ell}{ds} \quad . \tag{7.28}$$

Clearly the normal component could be due to ordinary 4D forces (it obeys $N^\mu u_\mu = 0$ by construction), *but there exists no 4D analog of the parallel component* (it is $P^\mu = u^\mu F^\alpha u_\alpha$ and has $P^\mu u_\mu \neq 0$ in general). This anomalous 'fifth force' is therefore a consequence of the existence of the extra dimension.

7.4. The Fifth Force

In the preceding section, we saw that when the metric has the general canonical form (7.1), the equations of motion in spacetime (7.19) consist of the standard 4D geodesics modified by a force (per unit mass) or acceleration which owes its existence to the fifth dimension. We arrived at this conclusion because the component of the new force parallel to the 4-velocity (7.28) has no analog in general relativity. Also, that force is inertial in the Einstein sense: It is proportional to the coupling between the 4D frame and the extra dimension $(\partial g_{\alpha\beta} / \partial \ell)$ and to the relative velocity $(d\ell/ds)$. We infer that the force depends on how the metric tensor varies with $x^4 = \ell$, and on the fact that we have used the 4D proper time s as parameter. These choices are natural but not unique (see below). For reasons given above, we prefer to keep s as parameter; but this leaves the question: Do we obtain the same expression for the fifth force if we alter the form of the metric, or equivalently the frame of coordinates?

This is of practical importance. The canonical form (7.1) consists of a quadratic function $(\ell/L)^2$ multiplied onto spacetime $g_{\alpha\beta}(x^\gamma, \ell)$ plus an extra flat piece. As we pointed out, it is *this* metric tensor which we employed in the normalization condition $g_{\alpha\beta}u^\alpha u^\beta = 1$ for the 4-velocities. But there are many exact solutions of the field equations in the literature which do not have metrics of canonical form; and anyhow we would like to know what the fifth force looks like for a general, non-factorized metric.

Let us suppose that we perturb the 4D part of a 5D metric, but instead of the canonical form we take the 4D line element in the more general form

$$ds^2 = g_{\alpha\beta}(x^\mu, \ell)\, dx^\alpha dx^\beta \quad . \tag{7.29}$$

Now we use *this* (unfactorized) metric tensor to normalize the 4-velocities; and to evaluate the parallel component of the fifth force we simply differentiate

$$g_{\alpha\beta}(x^\mu, \ell)u^\alpha u^\beta = 1 \tag{7.30}$$

with respect to proper time s. Hence

$$\partial_\gamma g_{\alpha\beta} u^\gamma u^\alpha u^\beta + \frac{\partial g_{\alpha\beta}}{\partial \ell}\frac{d\ell}{ds} u^\alpha u^\beta + 2g_{\alpha\mu}\frac{du^\mu}{ds}u^\alpha = 0 \quad , \tag{7.31}$$

where the first term on the left-hand side can be written as

$$(\partial_\gamma g_{\alpha\beta} + \partial_\beta g_{\alpha\gamma} - \partial_\alpha g_{\beta\gamma})u^\gamma u^\alpha u^\beta = 2g_{\alpha\mu}\Gamma^\mu_{\beta\gamma}u^\gamma u^\alpha u^\beta , \tag{7.32}$$

using symmetries under the exchange of α and β. It follows that Eq. (7.31) can be expressed as

$$2g_{\alpha\mu}u^\alpha \left(\frac{du^\mu}{ds} + \Gamma^\mu_{\beta\gamma} u^\beta u^\gamma\right) + \frac{\partial g_{\alpha\beta}}{\partial \ell}\frac{d\ell}{ds}u^\alpha u^\beta = 0 \quad . \tag{7.33}$$

This can be rewritten using a force per unit inertial mass:

$$F^\mu u_\mu = -\frac{1}{2}\frac{\partial g_{\alpha\beta}}{\partial \ell}\frac{d\ell}{ds}u^\alpha u^\beta \quad . \tag{7.34}$$

Thus in the new perturbed coordinates a timelike fifth force appears which is given by

$$P^\mu = \left(-\frac{1}{2}\frac{\partial g_{\alpha\beta}}{\partial \ell}u^\alpha u^\beta\right)\frac{d\ell}{ds}u^\mu \quad . \tag{7.35}$$

This is the same expression as Eq. (7.28). However, the nature of the argument presented here implies that the fifth force has this form *generally*, regardless of the coordinate system employed.

Experiments to detect various kinds of fifth force have been made through the years, directed at possible interaction in addition to the usual four (gravity, electromagnetism and the weak and strong nuclear forces). Our fifth force is somewhat different in nature, but the absence of any departure from 4D dynamics implies that it, too, must be small. By Eq. (7.35), the likely explanation for this is that the velocity $d\ell/ds$ is small, at least in laboratory situations. Also, any dynamical test for the fifth force P^μ in Eq. (7.35) has to contend with the presence of other forces normal to the 4-velocity N^μ, which are due to ordinary interactions.

What kind of effect should we look for in order to detect the fifth force? To answer this, consider a test particle of rest mass m, whose motion changes according to a generic form of Newton's second law:

$$\frac{D}{ds}(mu^\alpha) = m(N^\alpha + P^\alpha) \quad . \tag{7.36}$$

Using $u_\alpha u^\alpha = 1$ and $u_\alpha Du^\alpha/ds = 0$, we obtain

$$\frac{D}{ds}u^\alpha = N^\alpha \quad ,$$

$$P^\alpha = \frac{1}{m}\left(\frac{dm}{ds}\right)u^\alpha \quad . \tag{7.37}$$

This result shows that the force parallel to the 4-velocity, typical of 5D, would show up as a *variation of mass* in 4D.

The possibility that the mass of an object may vary was appreciated by Newton, who realized that force is not only proportional to acceleration but is properly defined as the rate of change of momentum. The geodesics of general relativity involve acceleration, but the motions of real objects are only described accurately in that way if their masses are constants. People tend to forget this. However, when we watch a rocket lift from a launch pad, we are basically seeing the acceleration which follows as it loses mass in accordance with the law of conservation of (linear) momentum. The latter, in its local form with curvature (gravitation) neglected, reads $d(mu^\alpha)/ds = 0$. That is,

$$\frac{du^\mu}{ds} + \frac{u^\mu}{m}\frac{dm}{ds} = 0 \quad . \tag{7.38}$$

Let us now assume that the acceleration is caused only by the fifth force as described by Eq. (7.35). This will enable us to see the effect of that force on the mass, and gives

$$\frac{1}{m}\frac{dm}{ds} = \left(\frac{1}{2}\frac{\partial g_{\alpha\beta}}{\partial \ell}u^\alpha u^\beta\right)\frac{d\ell}{ds} \quad . \tag{7.39}$$

Here, the velocity $d\ell/ds$ is inherently reversible (there is no friction associated with the fifth dimension). Taking account of this by a sign choice, and introducing an appropriate fiducial value of the mass, the integration of (7.39) yields

$$m = m_*\exp\left[\pm \int \left(\frac{1}{2}\frac{\partial g_{\alpha\beta}}{\partial \ell}u^\alpha u^\beta\right)d\ell\right] \quad . \tag{7.40}$$

This is an explicit expression for the rest mass of a test particle, assuming it is acted upon solely by the fifth force. It is in effect a definition for m, based on the fifth force and the law of conservation of momentum.

To make contact with Section 7.2, let us revert to canonical space by putting $g_{\alpha\beta}(x^\gamma, \ell) = (\ell/L)^2 \bar{g}_{\alpha\beta}(x^\gamma$ only) in Eq. (7.40). Then the coupling factor is $\frac{1}{2}(\partial g_{\alpha\beta}/\partial\ell)u^\alpha u^\beta = 1/\ell$, and is independent of the constant length L. We find that Eq. (7.40) gives

$$\frac{m}{m_*} = \left(\frac{\ell}{\ell_*}\right)^{\pm 1} \quad . \tag{7.41}$$

The integration constants have no special significance, while the sign choice merely reflects the reversibility of the motion in the extra dimension. The true significance of Eq. (7.41) is that *particle mass is related to the extra coordinate of* 5D *relativity*. With the constants made explicit, the interpretations of $x^4 = \ell$ implied by Eq. (7.41) are the $\ell_E = Gm/c^2$ and $\ell_P = h/mc$ introduced before.

The above concerns the special case of pure-canonical space, where the 5D metric has the form $dS^2 = g_{\alpha\beta}(x^\gamma, \ell)dx^\alpha dx^\beta - d\ell$ with $g_{\alpha\beta} = (\ell/L)^2 \bar{g}_{\alpha\beta}(x^\gamma)$, and we are normalizing the 4-velocities with $g_{\alpha\beta}$. It is important to realize that had we instead normalized the velocities with $\bar{g}_{\alpha\beta}$, we would have obtained the expression (7.19) of Section 7.2. While the two approaches are compatible, they use in effect different sets of coordinates. The canonical coordinates of the preceding section cause the $(\ell/L)^2$ factor to be decoupled from the 4-velocities and the rest of the dynamics, with the result that in Eq. (7.19) there is *no* fifth force if there is no additional dependence of the metric on ℓ, so the motion is 4D geodesic and there is no variation in mass. By contrast, the coordinates used in the present section are general, with the result that the fifth force defined by Eq. (7.35) is prevalent, and by the conservation of momentum there is a variation in mass, as given by Eq. (7.41) in mass. The message is one we have met before: The mass of a particle is a 4D quantity, and its behaviour depends on the 5D coordinate frame, which involves $x^4 = \ell$ in a significant way.

The comments above may appear obvious, especially in retrospect. But they help to dispel the confusion which surrounds the definition of particle mass in 5D relativity. In fact, it appears that there is no such thing as a unique definition. Thus, Ponce de Leon has argued for a restriction of the allowed groups of coordinate transformations, so as to give back 4D covariant dynamics [11]; while Bejancu *et al.* have argued for a kind of decomposition of a foliation into orthogonal components with their associated dynamical quantities, of the kind first proposed by Einstein and Bergmann [12]. However, these proposals are classical in nature, and do not take into account input from particle physics, which suggests that the Kaluza-Klein scalar field may be related to the Higgs field for mass, and that supersymmetry may be realizable if massive particles in 4D are photon-like in 5D [13]. In this regard, it was noted at the beginning of Section 7.3 that the most straightforward approach to 5D dynamics is via the null path $dS^2 = 0$. If the metric is complicated, it may be necessary to augment this by extra relations, but then these should be taken around the null path. It may be verified that the equation of motion in the extra dimension (7.20) holds in this context. Certainly for the approach to 5D physics of STM theory, the most logical postulate for the dynamics is $dS^2 = 0$.

7.5. The One-Body Problem

Campbell's theorem ensures, as we have seen, that the (pure) canonical metric includes all solutions of the Einstein equations with vacuum, including those of Schwarzschild and de Sitter [1, 14]. However, it is instructive to look at the 5D field equations for these cases, especially to verify that the Einstein vacuum is unique.

Consider therefore the static, 3D spherically-symmetric metric

$$dS^2 = \frac{\ell^2}{L_1^2}\left\{\left[1 - \frac{2M}{r} - \frac{r^2}{L_2^2}\right]dt^2 - \left[1 - \frac{2M}{r} - \frac{r^2}{L_2^2}\right]^{-1}dr^2\right.$$
$$\left. -r^2(d\Omega^2)\right\} - d\ell^2 \quad . \tag{7.42}$$

Here M is a constant associated with the mass at the centre of the 3-geometry, and L_1 and L_2 are constants with the dimensions of lengths. The non-zero components of R_{AB} for Eq. (7.42) are

$$R_{00} = \frac{3\left(rL_2^2 - 2ML_2^2 - r^3\right)\left(L_1^2 - L_2^2\right)}{rL_1^2 L_2^4} \quad ,$$

$$R_{11} = \frac{-3r\left(L_1^2 - L_2^2\right)}{L_1^2\left(rL_2^2 - 2ML_2^2 - r^3\right)} \quad ,$$

$$\tag{7.43}$$

$$R_{22} = \frac{-3r^2\left(L_1^2 - L_2^2\right)}{L_1^2 L_2^2} \quad ,$$

$$R_{33} = R_{22}\sin^2\theta \quad .$$

We see that the field equations $R_{AB} = 0$ are only satisfied if $L_1 = L_2 = L$ (say). Writing $L^2 = 3/\Lambda$ where Λ will be identified with the cosmological constant, Eq. (7.42) becomes

$$dS^2 = \frac{\Lambda \ell^2}{3}\left\{\left[1 - \frac{2M}{r} - \frac{\Lambda r^2}{3}\right]dt^2 - \left[1 - \frac{2M}{r} - \frac{\Lambda r^2}{3}\right]^{-1} dr^2 \right.$$
$$\left. - r^2(d\Omega^2) \right\} - d\ell^2 \quad . \tag{7.44}$$

The part of this metric inside the curly brackets is the 4D Schwarzschild-de Sitter one familiar from general relativity. The canonical solution (7.44) is quite different from the solitons discussed in Chapter 5.

The question arises of whether the 5D form (7.44) agrees with the classical tests in the solar system. This is easy to answer using the equations of motion (7.19). With the metric as set up in Section 7.3, $\partial g_{\alpha\beta}/\partial\ell = 0$ so the fifth force F^μ is zero. This means that the motion in 4D is identical to what it is in Einstein's theory. It follows that there is *no way* to tell by the classical tests of relativity if the solar system is described by the 5D metric (7.44) or its 4D part.

7.6. The Inflationary Universe

The 5D analogs of the standard 4D cosmologies with matter were examined in Section 3.2; but the very early universe is believed to have experienced a vacuum-dominated period of inflation described by an expanding de Sitter metric. To investigate this in the same manner as above, consider the 3D homogeneous, isotropic metric

$$dS^2 = \frac{\ell^2}{L_1^2}\left[dt^2 - e^{t/L_2}(dr^2 + r^2 d\Omega^2)\right] - d\ell^2 \quad . \tag{7.45}$$

The non-zero components of R_{AB} for this are

$$R_{00} = \frac{3(L_1^2 - 4L_2^2)}{4L_1^2 L_2^2} \quad ,$$

$$R_{11} = \frac{-3e^{t/L_2}(L_1^2 - 4L_2^2)}{4L_1^2 L_2^2} \quad ,$$

$$R_{22} = \frac{-3r^2 e^{t/L_2}(L_1^2 - 4L_2^2)}{4L_1^2 L_2^2} \quad , \tag{7.46}$$

$$R_{33} = R_{22}\sin^2\theta \quad .$$

We see that the field equations are only satisfied if $L_1 = 2L_2 = L$ (say). Writing again $L^2 = 3/\Lambda$, Eq. (7.45) becomes

$$dS^2 = \frac{\Lambda \ell^2}{3}\left[dt^2 - \exp\left(\sqrt{\frac{4\Lambda}{3}}t\right)(dr^2 + r^2 d\Omega^2)\right] - d\ell^2 \quad . \quad (7.47)$$

The part of this metric inside the square brackets is the 4D space-flat de Sitter one. The 5D solution (7.47) has $\partial g_{\alpha\beta}/\partial \ell = 0$ so $F^\mu = 0$ by Eq. (7.19). It follows again that there is no way using conventional dynamics to tell if inflationary cosmology is described by the 5D metric (7.47) or only its 4D part.

The results of this and the preceding section indicate that, for pure-canonical cases, traditional methods of probing for an extra dimension are not likely to succeed. It transpires, however, that the introduction of *spin* lifts the dynamical degeneracy noted above.

7.7. Spin and Precession

Since spin has the potential to differentiate in a testable way between 4D and 5D, we will in this section follow Liu and Wesson [15] and give a detailed account of the motion of a spinning object in the Schwarzschild-de Sitter metric.

We start with the pure-canonical 1-body solution, given by Eq. (7.44) in the form

$$dS^2 = \frac{\ell^2}{L^2}\left[\left(1 - \frac{2M}{r} - \frac{r^2}{L^2}\right)dt^2 - \left(1 - \frac{2M}{r} - \frac{r^2}{L^2}\right)^{-1}dr^2 - r^2 d\Omega^2\right] - d\ell^2 \quad . \quad (7.48)$$

Here L and M are constants. The 5D Christoffel symbols of the second kind for Eq. (7.48) are:

$$\Gamma_{tt}^{\ell} = -\frac{\ell(r^3 - rL^2 + 2ML^2)}{rL^4} \quad ,$$

$$\Gamma^\ell_{rr} = -\frac{\ell r}{(r^3 - rL^2 + 2ML^2)} \quad , \quad \Gamma^\ell_{\theta\theta} = -\frac{\ell r^2}{L^2} \quad ,$$

$$\Gamma^\ell_{\phi\phi} = \frac{\ell r^2(\cos 2\theta - 1)}{2L^2} \quad ,$$

$$\Gamma^t_{tr} = -\frac{(r^3 - ML^2)}{r(r^3 - rL^2 + 2ML^2)} \quad , \quad \Gamma^t_{t\ell} = -\frac{1}{\ell} \quad ,$$

$$\Gamma^r_{tt} = \frac{(r^3 - rL^2 + 2ML^2)(r^3 - ML^2)}{r^3 L^4} \quad ,$$

$$\Gamma^r_{rr} = -\frac{(r^3 - ML^2)}{r(r^3 - rL^2 + 2ML^2)} \quad , \quad \Gamma^r_{r\ell} = \frac{1}{\ell} \quad ,$$

$$\Gamma^r_{\theta\theta} = \frac{(r^3 - rL^2 + 2ML^2)}{L^2} \quad ,$$

$$\Gamma^r_{\phi\phi} = -\frac{(\cos 2\theta - 1)(r^3 - rL^2 + 2ML^2)}{2L^2} \quad ,$$

$$\Gamma^\theta_{r\theta} = \frac{1}{r} \quad , \quad \Gamma^\theta_{\theta\ell} = \frac{1}{\ell} \quad , \quad \Gamma^\theta_{\phi\phi} = -\frac{\sin 2\theta}{2} \quad ,$$

$$\Gamma^\phi_{r\phi} = \frac{1}{r} \quad , \quad \Gamma^\phi_{\theta\phi} = \cot\theta \quad , \quad \Gamma^\phi_{\phi\ell} = \frac{1}{\ell} \quad .$$

(7.49)

We will need these in order to carry out the rest of the analysis.

An object which moves on an orbit with a 5-velocity $U^A \equiv dx^A/dS$ and has an intrinsic spin vector S^A is governed by equations that reflect geodesic motion and relations set by the metric. For the orbit and spin, respectively, we have to solve

$$\frac{dU^A}{dS} + \Gamma^A_{BC} U^B U^C = 0 \quad , \quad g_{BC} U^B U^C = 1 \quad , \tag{7.50}$$

$$\frac{dS^A}{dS} + \Gamma^A_{BC} S^B U^C = 0 \quad , \quad g_{BC} S^B U^C = 0 \quad . \tag{7.51}$$

The orbit we choose to be circular and to lie in the plane $\theta = \pi/2$, so that $U^1 = U^r = 0$ and $U^2 = U^\theta = 0$. The remaining three components of the 5-velocity may be verified by substitution into Eq. (7.50) to be given by

$$U^0 = \frac{dt}{ds} = \frac{1}{\sqrt{1 - 3M/r}} \frac{L\ell_0}{\ell^2} \quad ,$$

$$U^3 = \frac{d\phi}{ds} = \frac{\sqrt{M/r - r^2/L^2}}{\sqrt{1 - 3M/r}} \frac{L\ell_0}{\ell^2 r} \quad , \tag{7.52}$$

$$U^4 = \frac{d\ell}{ds} = -\frac{S}{\ell} \quad .$$

Here ℓ_0 is an arbitrary constant. We leave the spin unrestricted for now, and its five components may be verified by a lengthy substitution into Eqs. (7.51) to be given by

$$S^0 = \frac{A_0}{\ell} \sin[\phi_0 - \Omega(s - s_0)] - \frac{B_0 S}{\ell^2} \quad ,$$

$$S^1 = \frac{A_1}{\ell} \cos[\phi_0 - \Omega(s - s_0)] \quad ,$$

$$S^2 = \frac{B_2}{\ell} \quad , \tag{7.53}$$

$$S^3 = \frac{A_3}{\ell} \sin[\phi_0 - \Omega(s - s_0)] - \frac{B_3 S}{\ell^2} \quad ,$$

$$S^4 = \frac{B_4}{\ell} \quad .$$

Here ϕ_0 and s_0 are arbitrary constants, and one should be careful to distinguish between the 5D scalar interval S, the 5D components of the spin vector S^A and the 4D scalar interval (or proper time) s. Also in Eqs. (7.53), we have introduced functions A_0, A_1, A_3 and B_0, B_2, B_3, B_4 which depend on r but on only three constants (say H_1, H_2, H_4):

$$A_0 = \frac{\sqrt{M/r - r^2/L^2}}{\sqrt{1 - 2M/r - r^2/L^2}} \frac{H_1 L}{\sqrt{1 - 3M/r}} \quad ,$$

$$A_1 = \sqrt{1 - 2M/r - r^2/L^2} \, H_1 L \quad ,$$

$$A_3 = \frac{\sqrt{1 - 2M/r - r^2/L^2}}{r \sqrt{1 - 3M/r}} H_1 L \quad ,$$

$$B_0 = \frac{H_4 L}{\sqrt{1-3M/r}} \quad , \tag{7.54}$$

$$B_2 = \frac{H_2 L}{r} \quad ,$$

$$B_3 = \frac{\sqrt{M/r - r^2/L^2}}{r\sqrt{1-3M/r}} H_4 L \quad ,$$

$$B_4 = H_4 \ell_0 \quad .$$

The constants H_1, H_2, H_4 are arbitrary but normalized amplitudes of the spin S^A, since the components of Eqs. (7.53) satisfy

$$S_A S^A = -(H_1^2 + H_2^2 + H_4^2) \quad . \tag{7.55}$$

By comparison, the parameter Ω which figures in Eqs. (7.53) is not arbitrary, and is in fact the spin angular velocity of the particle in its orbit:

$$\Omega = \frac{1}{r}\sqrt{\frac{M}{r} - \frac{r^2}{L^2}} \quad . \tag{7.56}$$

It should be noted that the latter is not the same as the orbital angular velocity, which is defined by

$$\omega \equiv \frac{d\phi}{ds} = \frac{1}{r}\frac{\sqrt{M/r - r^2/L^2}}{\sqrt{1-3M/r}} \quad . \tag{7.57}$$

The two are related by

$$\omega = \frac{\Omega}{\sqrt{1-3M/r}} \quad , \tag{7.58}$$

a discrepancy which produces the (geodetic) precession of the spin axis of an object in orbit around a central mass, even in 4D. We proceed to look at this for a reasonably physical situation.

Let us assume that a spinning object is put into circular orbit around a central mass so that its spin vector lies in the plane of the orbit.

Then $S^2 = 0$. Let us also assume that by mechanical means we have $rS^3 \ll S^1$ (due to our choice of coordinates S^0, S^1 and S^4 are dimensionless while S^2 and S^3 have dimension of inverse length, hence this physical condition). We make observations in 4D, so by analogy with what we did in previous chapters for the 4-velocities, we can define $S_s^A = S^A(dS/ds)$ where $dS/ds = \ell^2/L \ell_0$ by previous relations. Then by Eqs. (7.53) and (7.54), the components of S_s^A are given by:

$$S_s^0 = \frac{H_1 \ell}{l_0} \frac{\sqrt{M/r - r^2/L^2}}{\sqrt{(1-2M/r-r^2/L^2)(1-3M/r)}} \sin[\phi_0 - \Omega(s - s_0)]$$
$$- \frac{H_4 S}{\ell_0 \sqrt{1-3M/r}} \quad,$$

$$S_s^1 = \frac{H_1 \ell}{\ell_0} \sqrt{1 - \frac{2M}{r} - \frac{r^2}{L^2}} \cos[\phi_0 - \Omega(s - s_0)] \quad,$$

$$S_s^2 = \frac{H_2 \ell}{\ell_0 r} = 0 \quad, \tag{7.59}$$

$$S_s^3 = \frac{H_1 \ell}{\ell_0 r} \frac{\sqrt{1-2M/r-r^2/L^2}}{\sqrt{(1-3M/r)}} \sin[\phi_0 - \Omega(s - s_0)]$$
$$- \frac{H_4 S}{\ell_0 r} \frac{\sqrt{M/r-r^2/L^2}}{\sqrt{1-3M/r}} \quad,$$

$$S_s^4 = \frac{H_4 \ell}{L} \quad.$$

At the start of an orbit we set $s = s_0, S = S_0, \phi_0 = 0$. At the end of a single orbit $s = s_0 + \Delta s, S = S_0 + \Delta S$ where $\Delta s = 2\pi/\omega = 2\pi r \sqrt{1 - 3M/r}/\sqrt{M/r - r^2/L^2}$ by Eq. (7.57) and $\Delta S = (\ell^2/L\ell_0)\Delta s$ from above. The angle ϕ does not go exactly to 2π, but by analogy with the 4D problem has an offset $\delta\phi = r\Delta S_s^3/S_s^1$ where $\Delta S_s^3 \equiv S_s^3(s) - S_s^3(s_0)$. These relations combined with Eqs. (7.59) in the weak-field approximation $(r^2/L^2 \ll M/r \ll 1)$ give the precession angle for one orbit, and the effect will accumulate over $1/\Delta s \approx \sqrt{M/4\pi^2 r^3}$ orbits in a year. For an experiment of short duration $(\Delta s \ll L)$, the expression for the precession per orbit is fairly simple:

$$\delta\phi = \frac{3\pi M}{r} - 2\pi \left(\frac{H_4}{H_1}\right)\left(\frac{r}{L}\right) . \tag{7.60}$$

The first term here is the usual geodetic precession found in 4D general relativity. The second term will be small because it depends on the (presumably cosmological) length L in the metric. However, the terms in Eq. (7.60) depend quite differently on the radial distance r, so they could be separated, at least in principle.

Detecting the anomalous part of the precession in Eq. (7.60) is unfortunately difficult in practice. The Gravity Probe-B experiment, launched into low-earth orbit in 2004, was designed to directly test the existence of geodetic precession (as well as other, smaller predictions of general relativity) through a comparison of the spin axes of several supercooled spinning spheres with the line of sight to a distant guide star under drag-free conditions. The data confirmed the general relativistic prediction for geodetic precession to better than 0.3% [16]. This was sufficient to put useful constraints on some alternative theories of gravity [17], but not to detect an extra dimension [18]. To do so unequivocally would require the application of the preceding analysis to a system with much stronger gravitational fields, such as a binary pulsar.

7.8. Conclusion

Canonical space is of importance for physics on the large and small scales, because the universe appears to be dominated by the cosmological constant and particles are embedded in a sea of vacuum energy. In the general 5D metric, spacetime is factorized by a quadratic in the extra coordinate, and the extra dimension is flat. If the 4D part depends further on the extra coordinate there is matter present; while if it does not there is only vacuum (or equivalently a cosmological constant). If a shift is applied to the extra coordinate, a hypersurface appears where the cosmological 'constant' diverges (Section 7.2). In general, there is a fifth force due to the extra dimension, but it disappears for the pure-canonical metric when the coordinates are chosen so as to decouple spacetime (Sections 7.3, 7.4). As a consequence, there is no change for the dynamics of the 1-body problem and the inflationary universe (Sections 7.5, 7.6). This degeneracy is lifted if there is spin (Section 7.7). Even so, it is difficult to conceive of a practical test for the fifth dimension.

The current paradigm for constructing alternative theories of gravity is that they should reduce smoothly to general relativity. This is understandable; but then it follows that any new effects are small and likely hard to detect. Given this situation, it is natural to look in new directions for evidence of the fifth dimension.

References

[1] Mashhoon, B., Liu, H. and Wesson, P.S., *Phys. Lett.* **B331**, 305 (1994).

[2] Rindler, W., *Essential Relativity*, 2nd. ed. (Springer, New York, 1977), p. 185.

[3] Mashhoon, B. and Wesson, P.S., *Class. Quant. Grav.* **21**, 3611 (2004); *Gen. Rel. Grav.* **39**, 1403 (2007).

[4] Ponce de Leon., J., *Grav. Cosmol.* **14**, 241 (2008).

[5] Wald, R.M., *General Relativity* (Un. Chicago Press, Chicago, 1984), p. 446.

[6] Seahra, S.S. and Wesson, P.S., *Gen. Rel. Grav.* **33**, 1731 (2001).

[7] Youm, D., *Mod. Phys. Lett.* **A16**, 2371 (2001).

[8] Mashhoon, B., Wesson, P.S. and Liu, H., *Gen. Rel. Grav.* **30**, 555 (1998).

[9] Wesson, P.S., Mashhoon, B., Liu, H. and Sajko, W.N., *Phys. Lett.* **B456**, 34 (1999).

[10] Youm, D., *Phys. Rev.* **D62**, 084002 (2000).

[11] Ponce de Leon, J., *Phys. Lett.* **B523**, 34 (2001). *Grav. Cosmol.* **8**, 272 (2002); *Int. J. Mod. Phys.* **D12**, 757 (2003).

[12] Einstein, A. and Bergmann, P., *Ann. Math.* **39**, 683 (1938); Bejancu, A., Călin, C. and Farran, H.R., *J. Math. Phys.* **53**, 122503 (2012).

[13] Dawson, S., Brookhaven National Lab preprint BNL-HET-99/ and arXiv preprint hep-ph/9901280 (1999); Bettini, A., *Introduction to Elementary Particle Physics* (Cambridge Un. Press, Cambridge, 2008).

[14] Rippl, S., Romero, C. and Tavakol, R., *Class. Quant. Grav.* **12**, 2411 (1995); Romero, C., Tavakol, R. and Zalaletdinov, R., *Gen. Rel. Grav.* **28**, 365 (1996); Lidsey, J.E., Romero, C., Tavakol and R., Rippl, S., *Class. Quant. Grav.* **14**, 865 (1997). See also Fonseca-Neto, J.B. and Romero, C., *Class. Quant. Grav.* **24**, 3515 (2007). This discusses an apparently new embedding, which however is really a coordinate transformation where $x^4 = \ell$ is shifted by a constant.

[15] Liu, H. and Wesson, P.S., *Class. Quant. Grav.* **13**, 2311 (1996).

[16] Everitt, C.W.F. *et al.*, *Class. Quant. Grav.* **32**, 224001 (2015).

[17] Overduin, J.M., *Class. Quant. Grav.* **32**, 224003 (2015).

[18] Overduin, J.M., Everett, R.D. and Wesson, P.S., *Gen. Rel. Grav.* **45**, 1723 (2013).

Chapter 8

Particles, Waves and Vacuum

8.1. Introduction

In principle, a field theory which is broad enough can be applied to physics on the large scale and the small scale. This is the idea behind using a higher-dimensional space to unify classical gravity with the quantum interactions of particles. Space-Time-Matter theory is like this. It uses five dimensions to unify the tensor, vector and scalar interactions which have to do with gravity, electromagnetism and matter. The last has been studied in previous chapters mainly in regard to fluids, but in the present chapter the emphasis is on particles. These can also, of course, manifest themselves as waves; and it will be seen that the unifying agent of wave-particle duality is the vacuum.

The idea that local physics is connected to global physics is embodied in Mach's Principle. We will give a brief summary of this, and attempt to make it more concrete than hitherto, in Section 8.2. Then in Section 8.3 we will discuss those properties of particles which can be fit naturally into the 5D theory. It is possible, in fact, to develop a formal scheme which relates the rest mass of a particle to the 4D curvature of the space it inhabits, as shown in Section 8.4. Then in Section 8.5 we will see that the duality between the momentum of a particle and its de Broglie wavelength can be understood as a consequence of 5D canonical space and the underlying vacuum. In Section 8.6 canonical space will also be employed to construct a model of a quantized wavicle. Section 8.7 is a conclusion. The philosophy throughout this chapter is pragmatic: in the absence of a quantum theory of gravity, or a higher-dimensional theory which properly accounts for the internal symmetry groups of particles, it seems prudent to aim for a theory that encompasses general relativity and wave mechanics.

8.2. Mach's Principle

Most principles of physics are ambiguous when expressed in words, and only gain precision and traction in the form of equations. The reason why Mach's Principle has influenced scientists since Einstein is that it embodies the vague dissatisfaction they feel about the nature and definition of the rest mass of a particle: is it merely a local quantity, or connected in some way to the other matter in the universe?

Newton's second law tells us that the mass m of a particle is the proportionality factor between the force and the acceleration. As such, it measures the resistance to motion, or inertia. Something similar is involved in what is widely called Newton's bucket experiment. A bucket full of water is suspended by a twisted rope, which is allowed to unwind. The bucket acquires a rotation, but the water inside remains flat and still (until friction between the bucket and its contents upsets things). Why the difference in the motions? We categorize the reluctance of the water to move by saying it has inertia. But where does this originate? We might reasonably expect the water to revolve immediately with the bucket, if the physics were purely local. The fact that this does not happen suggests that the water's stationary state, and inertia, are the result of some interaction that is long-distance or global.

Mach ruminated on the laws of mechanics, publishing his conclusions in a series of essays and in a much-read textbook [1]. His views influenced Einstein, who attempted to incorporate them into general relativity. For a while, there was hope that the Lense-Thirring effect – where a rotating massive shell drags spacetime around with it – might prove to be an inertial phenomenon of Machian type. Also known as 'frame-dragging,' this effect certainly exists: it was directly observed by Gravity Probe-B, the same experiment that detected geodetic precession

(Section 7.7). However, in the gravitational field of the rotating Earth, it is an exceedingly tiny effect – almost 200 times smaller than geodetic precession – and it was measured with a precision of only about 20%. Significant theoretical effort has been extended to determine whether frame-dragging fulfils Machian expectations when extrapolated to cosmological scales, but consensus on this issue remains elusive. Additionally, other aspects of general relativity run somewhat counter to Mach's ideas. The cosmological constant, for example, seems to endow empty space itself with tangible properties such as energy density and pressure, independent of interactions with matter.

Since the days of Mach and Einstein, there have been numerous attempts to modify the theory of the latter to incorporate the ideas of the former, but none has been generally accepted. Perhaps the best regarded of these accounts was that of Sciama, who outlined a theory where inertia is a field like gravity, but which varies with distance proportional to $1/r$ instead of $1/r^2$, so in a uniform universe the increase in the number of sources as r^3 means that the distant ones dominate the properties of a local particle. A more general proposition was that of Hoyle, who pointed out that Mach's Principle would be effectively realized (irrespective of the underlying interaction) if it were possible to write down an expression for the rest mass of a particle in the universe as a function of the spacetime coordinates. Reviews of Mach's Principle and related concepts of 4D physics show that different workers have very different ideas of how it should be implemented [2].

Reading Mach's original words shows that he was primarily concerned about the way in which the rest mass of a particle operates independently of the coordinates of space and time (and the underlying nature of spacetime). He preferred some scheme in which these things were integrated, so that m was organically connected to x^γ. He wrote: "Instead of referring a moving body to space, let us view directly its relation to the bodies of the universe, by which alone such a system of coordinates can be determined."

Previews are available which look at the foundations of physics generally and the formulation of Machian mechanics in particular [3, 4]. Here, we proceed to outline two ways in which physics can be formulated in the manner envisaged by Mach:

(i) Wave mechanics: This continues the mass m with the 4-velocities $u^\alpha \equiv dx^\alpha/ds$ and regards the 4-momenta $p^\alpha \equiv mu^\alpha$ as the prime data. De Broglie realized in 1923 that the p^α can be replaced by wavelengths, which in symbolic form are $\lambda^\alpha \sim 1/p^\alpha$ and are now named after him. These are derived from a complex wave function ψ, via $p_\alpha = (h/i\psi)(\partial\psi/\partial x^\alpha)$. Here h is Planck's constant, which is mathematically necessary to make dimensionless the exponent of the wave function for a single particle $\psi \sim e(ip_\alpha\psi^\alpha/h)$ and is physically the basis for the quantization implied by the waves. The wave function is a solution of the relativistic wave equation

$$\Box^2\psi + (mc/h)^2\psi = 0 \quad . \tag{8.1}$$

Here $\Box^2\psi \equiv g^{\alpha\beta}\nabla_\beta(\partial\psi/\partial x^\alpha)$ is the D'Alembertian for a spacetime with metric tensor $g_{\alpha\beta}$ for which the covariant (curved-space) derivative is indicated by a nabla symbol ("∇"), and there is summation over repeated indices as usual. The details may be familiar, but it is important to realize that the Klein-Gordon equation (8.1) is effectively the field equation for this system of mechanics. Irrespective of its background and limited scope, wave mechanics as just summarized is just what its name implies. And while the developers of wave mechanics may not have thought much about him, the theory does have some of the features advocated by Mach.

(ii) Space-Time-Matter theory: This was founded in 1992 and developed as a theory where the fields are united with the their sources, so it is to be expected that it is Machian in nature. But while it is guaranteed by Campbell's theorem that Einstein's 4D theory with fluid sources can be embedded in the 5D field equations $R_{AB} = 0$, there is still the question of how to handle an individual particle. The basic answer to this was given in Chapter 7, which dealt with 5D canonical space. The (pure) canonical metric embeds all solutions of Einstein's equations without ordinary matter but with a vacuum whose energy density is measured by the cosmological constant, thus:

$$dS^2 = (\ell/L)^2 g_{\alpha\beta}(x^\gamma)dx^\alpha dx^\beta \pm d\ell^2 \quad ,$$

$$\tag{8.2}$$

$$\Lambda = \mp 3/L^2 \quad .$$

This metric, it will be recalled, was introduced with the idea that the extra coordinate $x^4 = \ell$ would play the role of the mass m of an isolated object situated in vacuum [5]. Then the 4D part of the 5D metric gives back the action $m\,ds$ of conventional mechanics. The original context was gravitational, so the appropriate identification was in terms of the gravitational constant and the Schwarzschild radius, with $\ell = Gm/c^2$. The theory is covariant, however, so the coordinate transformation $\ell \to L^2/\ell$ brings in Planck's constant and the Compton wavelength, with the alternative identification $\ell = h/mc$. In other words, for the metric (8.2) the particle mass can be taken as the coordinate, making STM theory manifestly Machian.

The preceding accounts of wave mechanics and STM theory are cursory. But that these theories are similar at base may be stated in an even more succinct manner. For if we combine the de Broglie wavelengths with the Compton wavelength and form the corresponding wave numbers, then the 5D statement $K^A K_A = 0$ is essentially the same as saying that the 5D geodesic is null or $dS^2 = 0$. These are equivalent relations for the dynamics in different languages. In the remainder of this chapter, we proceed to delve into the connection between wave mechanics and STM theory in more detail.

8.3. Particle Properties and 5D

In previous chapters we obtained expressions for certain classical properties of particles based on 5D. For example, in Section 6.7 we used the standard electromagnetic form of the metric to derive the charge-to-mass ratio of a particle; and in Section 7.4 we used the canonical metric to determine the mass of an uncharged particle, finding in Eq. (7.41) that it is proportional to the extra coordinate. In the present section, we wish to probe somewhat deeper. After some comments about the origin of the Klein-Gordon equation (8.1) and its companion, the Dirac equation, we will inquire about the mass scales of particles, their quasi-quantum behaviour, and the relationship between particle mass and the cosmological constant.

The Klein-Gordon equation (8.1) was originally derived by applying 4D spin-0 wave operators to the normalization condition for the energy, momenta and mass of a particle:

$$E^2 - p^2 c^2 - m^2 c^4 = 0 \quad . \tag{8.3}$$

But as we have noted elsewhere, this condition is really nothing more than an expanded version of the 4-velocity condition $u^\alpha u_\alpha = 1$ commonly applied in special and general relativity. The mass m is introduced simply by multiplying the last relation by m^2, giving Eq. (8.3). We see that the Klein-Gordon equation, and its non-relativistic version, the Schrödinger equation, do not contain any useful information about the rest mass of a particle. A more meaningful derivation will be given in Section 8.5, where it will be shown that the Klein-Gordon equation is equivalent to the last component of the 5D geodesic equation.

The Dirac equation was derived by 'factorizing' the energy condition (8.3). It relates the mass m of a spin-1/2 particle to a bispinor field ψ via

$$i\gamma^\alpha \frac{\partial \psi}{\partial x^\alpha} - \frac{mc}{h} \psi = 0 \quad . \tag{8.4}$$

Here γ^α are four 4×4 matrices which obey the relationship $\gamma^\alpha \gamma^\beta + \gamma^\beta \gamma^\alpha = 2\eta^{\alpha\beta}$. This effectively decomposes the metric of spacetime, which is here taken to be given by the Minkowski form with $\eta^{\alpha\beta} = (+1 - 1 - 1 - 1)$. There is a considerable literature on the 5D version of Eq. (8.4). However, the starting point for a 5D derivation can be taken to be something both simple and familiar (see Section 7.7). Namely, the fact that the 5D spin vector S_A is spacelike while the ordinary velocity vector dx^A/ds is timelike, so their inner product can always be made to vanish [6]. From this, some algebra leads to a 5D form of the Dirac equation. It should be remarked that the applications of this equation can be expedited by using a 5D identity found by Hoyle and Narlikar, which with the usual terminology and $A = (\alpha, 4)$ can be written

$$\left| \bar{\psi}\psi \right|^2 \equiv (\bar{\psi}\gamma^\alpha\psi)(\bar{\psi}\gamma_\alpha\psi) - (\bar{\psi}\gamma_4\psi)^2 \quad . \tag{8.5}$$

This may be proved using the Pauli matrices. In application to 5D, the main thing to note is that the Dirac equation (8.4) and the Klein-Gordon equation (8.1) both use the Compton wavelength h/mc to measure the mass, and this is acceptable as a coordinate in certain 5D metrics.

The masses of elementary particles do not appear to follow any well-defined rule, or to be quantized in terms of some universal unit. However, if to the fundamental constants h, G and c we add the cosmological constant Λ with its astrophysically-determined value (namely, 1.1×10^{-52} m^{-2}), then we can form two unique masses [7]:

$$m_h \equiv \left(\frac{h}{c}\right)\left(\frac{\Lambda}{3}\right)^{\frac{1}{2}} = 1.3 \times 10^{-68} \text{ kg} \quad , \tag{8.6}$$

$$m_G \equiv \left(\frac{c^2}{G}\right)\left(\frac{3}{\Lambda}\right)^{\frac{1}{2}} = 2.2 \times 10^{53} \text{ kg} \quad . \tag{8.7}$$

These are relevant to quantum and gravitational systems, and should not be confused with the so-called Planck mass $m_p \equiv \sqrt{\hbar c/G} \approx 2 \times 10^{-8}$ kg, which mixes h and G but does not involve Λ. The second number above is roughly equal to the mass of the observable universe, and is equivalent to 10^{80} baryons of mass 1.7×10^{-27} kg each. The first number is more difficult to interpret, but appears to be the mass of a quantum perturbation in a universe with a very small average curvature.

A very small quantum of mass actually results from an analysis of the canonical metric if there is structure in the fifth dimension [8]. To see this, we take the pure-canonical metric of Chapter 7, do the coordinate transformation $L \to L^2/\ell$, and identify the extra coordinate with the Compton wavelength $\ell = h/mc$ (this means that the 4D part of the 5D metric contains the usual action $mc\ ds$). Then we obtain Eq. (7.7) or:

$$dS^2 = \frac{L^2}{\ell^2}ds^2 - \frac{L^4}{\ell^4}d\ell^2 \quad , \tag{8.8}$$

where the fifth dimension is taken to be spacelike. The Lagrangian density $\mathcal{L} = (dS/ds)^2$ corresponding to Eq. (8.8) has associated with it 5-

momenta given by

$$P_\alpha = \frac{\partial \mathcal{L}}{\partial(dx^\alpha/ds)} = \frac{2L^2}{\ell^2} g_{\alpha\beta} \frac{dx^\beta}{ds} \quad,$$

(8.9)

$$P_\ell = \frac{\partial \mathcal{L}}{\partial(d\ell/ds)} = -\frac{2L^4}{\ell^4} \frac{d\ell}{ds} \quad.$$

These define a 5D scalar which is the analog of the one used in 4D quantum mechanics:

$$\int P_A dx^A = \int (P_\alpha dx^\alpha + P_\ell d\ell)$$
$$= \int \frac{2L^2}{\ell^2}\left[1 - \left(\frac{L}{\ell}\frac{d\ell}{ds}\right)^2\right] ds \quad.$$

(8.10)

This is zero for $dS^2 = 0$, in which case Eq. (8.8) gives

$$\ell = \ell_* e^{\pm s/L} \quad, \quad \frac{d\ell}{ds} = \pm\frac{\ell}{L} \quad,$$

(8.11)

where ℓ_* is a constant. This is the same kind of behaviour found in previous applications of the canonical metric. The test particle we are considering has finite energy in 4D, but zero 'energy' in 5D because $\int P_A dx^A = 0$. The corresponding quantity in 4D is $\int p_\alpha dx^\alpha$, and using relations from above it is given by

$$\int p_\alpha dx^\alpha = \int mu_\alpha dx^\alpha = \int \frac{h\,ds}{c\ell} = \pm\frac{h}{c}\frac{L}{\ell} \quad.$$

(8.12)

The fact that this can be positive or negative goes back to Eq. (8.11), but since the motion is reversible we will suppress the sign in what follows for convenience. We will also put $L/\ell = n$, anticipating a physical interpretation which indicates that it is not only dimensionless but may be a rational number. Then Eq. (8.12) says

$$\int mc\,ds = nh \quad.$$

(8.13)

That is, the conventional quantization rule in 4D follows from a null line element in 5D.

The other scalar quantity that is of interest in this approach is $dp_\alpha dx^\alpha$. (It should be recalled that dx^α transforms as a tensor but x^α does not.) Following the same procedure as above, one finds

$$dp_\alpha dx^\alpha = \frac{h}{c}\left(\frac{du_\alpha}{ds}\frac{dx^\alpha}{ds} - \frac{1}{\ell}\frac{d\ell}{ds}\right)\frac{ds^2}{\ell} . \tag{8.14}$$

The first term inside the parentheses here is zero if the acceleration is zero, or if the scalar product with the velocity is zero as in conventional 4D dynamics (see Section 7.3). But even so, there is a contribution from the second term inside the parentheses which is due to the change in mass of the particle. This anomalous contribution has magnitude

$$|dp_\alpha dx^\alpha| = \frac{h}{c}\left|\frac{d\ell}{ds}\right|\frac{ds^2}{\ell^2} = \frac{h}{c}\frac{ds^2}{L\ell} = n\frac{h}{c}\left(\frac{d\ell}{\ell}\right)^2 , \tag{8.15}$$

where we have used Eq. (8.11) with $n = L/\ell$. The latter implies $dn/n = -d\ell/\ell = dK_\ell/K_\ell$ where $K_\ell \equiv 1/\ell$ is the wavenumber for the extra dimension. The last relation can be written

$$|dp_\alpha dx^\alpha| = \frac{h}{c}\frac{dn^2}{n} . \tag{8.16}$$

Thus a Heisenberg-type relation in 4D follows if $n = L/\ell$ and there is structure in the fifth dimension.

The last relation (8.16) is best interpreted by shifting to a wave picture, where the particle's momenta are given by its de Broglie wavelengths and the mass by its Compton wavelength. The de Broglie waves can have any size if the 4D topology is open, but are constrained if it is closed. For the latter case, we will look at a detailed model in Section 8.6. Here we note that the ratio of the scale of the geometry to the Compton wavelength determines $n = L/\ell$, which affects whether the scalar defined by Eq. (8.16) obeys the conventional uncertainty relation or not. The fundamental mode ($n = 1$) deserves special comment. This can be studied using Eqs. (8.13)–(8.16), or directly from Eq. (8.8) by using $\ell = h/mc$ with $dS^2 = 0$. The latter procedure gives $|dm| = mds/\ell$ which with Eq. (8.13) yields $m = (\int mcds)/cL = nh/cL$. This defines for $n = 1$ a fundamental unit mass h/cL. Since $\Lambda = 3/L^2$, this is $(h/c)(\Lambda/3)^{1/2}$ which is identical to Eq. (8.6) above.

The relationship between particle mass m and the cosmological constant Λ can be investigated in a more systematic way than above. We recall from Chapter 7 that for the 5D (pure) canonical metric, the form of Λ depends on the prefactor for the 4D part of the metric. Thus $\Lambda = 3/L^2$ when the prefactor is $(\ell/L)^2$, and $\Lambda = (3/L^2)l^2(l - l_0)^{-2}$ when the prefactor has a shift and is $(\ell - \ell_0)^2/\ell^2$. These prefactors, as far as the 4D part of the metric are concerned, are in the nature of *conformal* factors. It is well known how a conformal factor affects the Riemann-Christoffel tensor, the Ricci tensor and the Ricci or curvature scalar [9]. And using this, it is possible to work out how Λ and m scale under a conformal transformation $\Omega(\ell)$, subject to certain physical conditions. The latter, for us, may reasonably include the invariance of the classical action law $1 \equiv \int mc\, ds$ as we move from on ℓ-hypersurface to another. It transpires that under a conformal transformation by $\Omega(\ell)$, the metric tensor and Ricci scalar change to

$$g'_{\alpha\beta} = \Omega^2 g_{\alpha\beta} \quad , \quad R' = \Omega^{-2} R \quad . \tag{8.17}$$

Also, the cosmological constant and the particle mass change to

$$\Lambda' = \Omega^{-2}\Lambda \quad , \quad m' = \Omega^{-1} m \quad . \tag{8.18}$$

This implies that for manifolds whose physical parameters scale conformally, we could expect to find a relation of the form $\Lambda \propto m^2$.

8.4. Mass and Curvature

If Einstein had applied his geometrical approach to wave mechanics instead of gravitation, he might have come up with a relation of the form $(mc/h)^2 \propto R$ between the mass m of a particle and the scalar curvature of the 4D spacetime it inhabits. Such a relation basically equates the Compton wavelength h/mc to the radius of curvature $R^{-1/2}$. As an expression which connects the rest mass of a local particle to the global properties of spacetime (and its matter), it is eminently Machian. It is also in agreement with what we learned in the latter part of the preceding section about conformal factors (such a factor changes the metric tensor and so changes R, and in STM theory the factor may be related to m^2). In the present section, we wish to examine the consequences of applying conformal factors to both the 4D and extra parts of a 5D metric, particularly in regard to particle mass.

Wave mechanics, as developed by de Broglie, Schrödinger and others, replaces the momenta p^α of a particle by wavelengths which in symbolic fashion can be written $\lambda^\alpha = h/p_\alpha$. It is logical in a theory like STM to relate the λ^α to the principal radii of curvature of the spacetime, as described by the Ricci tensor $R_{\alpha\beta}$ or the Einstein tensor $G_{\alpha\beta} \equiv R_{\alpha\beta} - Rg_{\alpha\beta}/2$. And R or $G = -R$ is an obvious scalar to relate to the scalar mass m. This leads us to conjecture that the mass of a particle is related to the curvature of the space it inhabits by a relation of the form

$$m^2 = \frac{k\,h^2 R}{c^2} \ . \tag{8.19}$$

Here k is a dimensionless coupling constant, to be determined by a detailed analysis like that given below. It is important to appreciate that defining the mass via a relation like Eq. (8.19) is different from what we did before. Previously, we related m to the extra coordinate $x^4 = \ell$. Specifically, for the (pure) canonical metric of Chapter 7, we found that there is a special choice of coordinates which allows us to write $\ell = Gm/c^2$, or $\ell = h/mc$ if the coordinate is inverted. Now, in Eq. (8.19), there is no connection of the form $m = m(\ell)$; and in fact we will proceed with m as just a parameter, and a metric which for algebraic tractability we take to be independent of ℓ. The analysis follows Liu and Wesson [10], though their $\hbar$ is here replaced by h for uniformity.

Consider a metric of the form

$$dS^2 = g_{AB}dx^A dx^B = \Phi^2 ds^2 - \Phi^{2b}d\ell^2 \ ,$$

$$ds^2 \equiv g_{\alpha\beta}dx^\alpha dx^\beta \ . \tag{8.20}$$

This is fairly general, but we assume $\Phi = \Phi(x^\alpha)$ in order to make contact with the conventional wave function of quantum mechanics (see below). The Christoffel symbols for Eqs. (8.20) are

$$\hat{\Gamma}^\alpha_{\beta\gamma} = \Gamma^\alpha_{\beta\gamma} + \Phi^{-1}\left(\delta^\alpha_\beta \Phi_\gamma + \delta^\alpha_\gamma \Phi_\beta - g_{\beta\gamma}\Phi^\alpha\right) \quad ,$$

$$\hat{\Gamma}^\alpha_{\beta\alpha} = \Gamma^\alpha_{\beta\alpha} + 4\Phi^{-1}\Phi_\beta \quad ,$$

$$\Gamma^\alpha_{44} = b\Phi^{2b-3}\Phi^\alpha \quad ,$$

$$\Gamma^\alpha_{\beta 4} = \Gamma^4_{\alpha\beta} = \Gamma^4_{44} = 0 \quad , \qquad\qquad\qquad (8.21)$$

$$\Gamma^4_{4\alpha} = b\Phi^{-1}\Phi_\alpha \quad ,$$

$$\Gamma^D_{\alpha D} = \Gamma^\delta_{\alpha\delta} + (b+4)\Phi^{-1}\Phi_\alpha \quad ,$$

$$\Gamma^D_{4D} = 0 \quad .$$

Here as elsewhere $\Phi_\alpha \equiv \partial_\alpha\Phi$, $\Phi^\alpha = g^{\alpha\beta}\Phi_\beta$, a repeated index implies a sum, and hats denote the 4D parts of 5D quantities. The 5D Ricci tensor can be written in terms of the 4D Ricci tensor as usually defined plus other terms, thus:

$$\hat{R}_{\alpha\beta} = R_{\alpha\beta} - (b+2)\Phi^{-1}\nabla_\beta \Phi_\alpha - (b^2 - 3b - 4)\Phi^{-2}\Phi_\alpha\Phi_\beta$$
$$- \Phi^{-2}\left[\Phi\nabla_\gamma\Phi^\gamma + (b+1)\Phi^\gamma\Phi_\gamma\right]g_{\alpha\beta} \quad ,$$

$$R_{4\alpha} = 0 \quad , \qquad\qquad\qquad (8.22)$$

$$R_{44} = b\Phi^{2b-4}\left[\Phi\nabla_\alpha\Phi^\alpha + (b+1)\Phi^\alpha\Phi_\alpha\right] \quad .$$

Clearly the electromagnetic-type components of the field equations $R_{AB} = 0$ are trivially satisfied, while the remaining equations read

$$R_{\alpha\beta} = (b+2)\Phi^{-1}\nabla_\beta\Phi_\alpha + (b^2 - 3b - 4)\Phi^{-2}\Phi_\alpha\Phi_\beta \quad , \qquad (8.23)$$

$$\Phi\nabla_\alpha\Phi^\alpha + (b+1)\Phi^\alpha\Phi_\alpha = 0 \quad . \qquad\qquad\qquad (8.24)$$

These are a set of ten Einstein-like equations and one wave-like equation, and are what we will be concerned with in what follows.

From Eqs. (8.23) and (8.24) we find that the 4D Ricci scalar is

$$R = -6(b+1)\Phi^{-2}\Phi^{\alpha}\Phi_{\alpha} \ . \tag{8.25}$$

This combined with Eq. (8.23) allows us to form the Einstein tensor, and Einstein's equations $G_{\alpha\beta} = 8\pi T_{\alpha\beta}$ then define an induced energy-momentum tensor given by

$$8\pi T_{\alpha\beta} = (b+2)\Phi^{-1}\nabla_{\beta}\Phi_{\alpha} + (b+1)(b-4)\Phi^{-2}\Phi_{\alpha}\Phi_{\beta}$$
$$+3(b+1)\Phi^{-2}\Phi^{\gamma}\Phi_{\gamma}g_{\alpha\beta} \ . \tag{8.26}$$

This obviously contains a term that is second-order in the derivatives of the scalar field, and two terms that are first-order as in flat-space quantum field theory [11]. Because of this, we choose $b = -2$ at this point. This is also the choice that is compatible with the 5D harmonic coordinate condition, which implies wave-like solutions [12]. Then the 5D and 4D line elements are

$$dS^2 = \Phi^2 ds^2 - \Phi^{-4}d\ell^2 \ , $$
$$ds^2 \equiv g_{\alpha\beta}dx^{\alpha}dx^{\beta} \ , \tag{8.27}$$

where $g_{\alpha\beta} = g_{\alpha\beta}(x^{\gamma})$, $\Phi = \Phi(x^{\gamma})$. And the 5D field equations $R_{AB} = 0$ yield the 4D ones

$$R_{\alpha\beta} = 6\Phi^{-2}\Phi_{\alpha}\Phi_{\beta} \ , \tag{8.28}$$

$$\Phi\nabla_{\alpha}\Phi^{\alpha} - \Phi^{\alpha}\Phi_{\alpha} = 0 \ , \tag{8.29}$$

with an effective 4D energy-momentum tensor

$$8\pi T_{\alpha\beta} = 6\Phi^{-2}\left(\Phi_{\alpha}\Phi_{\beta} - \Phi^{\gamma}\Phi_{\gamma}g_{\alpha\beta}/2\right) \ . \tag{8.30}$$

These relations are expected to define the geometrical properties of a scalar wave which in terms of quantum theory would be described as a massive particle.

In Section 6.7, we reviewed the conventional way of describing the 4D dynamics of a particle in classical theory. For the situation here, the Lagrangian is $L = m(ds/d\lambda) = m[g_{\alpha\beta}(dx^\alpha/d\lambda)(dx^\beta/d\lambda)]^{1/2}$, where λ is a parameter for the path and the mass is assumed to be a constant. The momenta are $p_\alpha = \partial L/\partial(dx^\alpha/d\lambda) = mg_{\alpha\beta}(dx^\beta/ds)$, with $p^\alpha = g^{\alpha\beta}p_\beta = m(dx^\alpha/ds)$, so that $p_\alpha p^\alpha = m^2$. The action corresponding to the Lagrangian is $I = \int L\,d\lambda = \int m\,ds = \int m[g_{\alpha\beta}(dx^\alpha/ds)(dx^\beta/ds)]ds = \int p_\alpha\,dx^\alpha$. However, in place of the action we can use the wave function of quantum theory:

$$\Psi \equiv e^{-iI/\hbar} \quad . \tag{8.31}$$

In terms of this, the momenta are given by

$$i\hbar\frac{\partial}{\partial x^\alpha}\Psi = p_\alpha\Psi \quad . \tag{8.32}$$

If the p_α were constants, as in flat spacetime, they would be the eigenvalues and Ψ the eigenfunction of the 4D momentum operator $i\hbar\,\partial/\partial x^\alpha$. The p_α are not generally constants in curved spacetime, but we can keep Eq. (8.32) as the defining relation for the momenta because the preceding relations are covariant. Then following the philosophy of induced-matter theory, the question is how to relate the quantum relations (8.31), (8.32) in Ψ to the classical relations (8.28), (8.29) in Φ.

We answer this by adopting the ansatz

$$\Phi = \Psi^\varepsilon \quad , \tag{8.33}$$

where ε is a dimensionless arbitrary parameter. Then Eqs. (8.28) and (8.29) give

$$R_{\alpha\beta} = -\frac{6\varepsilon^2}{\hbar^2}p_\alpha p_\beta \quad , \tag{8.34}$$

$$\nabla_\alpha \Psi^\alpha + \frac{m^2}{h^2} \Psi = 0 \quad . \tag{8.35}$$

The corresponding induced energy-momentum tensor is, from Eq. (8.30):

$$8\pi T_{\alpha\beta} = -\frac{6\varepsilon^2}{h^2} \left(p_\alpha p_\beta - \frac{m^2 g_{\alpha\beta}}{2} \right) \quad . \tag{8.36}$$

These relations satisfy the picture we set out to paint: Eq. (8.34) says that the de Broglie wavelengths associated with p_α describe by their product $p_\alpha p_\beta$ the curvature of the 4D space as represented by the Ricci tensor $R_{\alpha\beta}$; Eq. (8.35) says there is a Klein-Gordon equation for a particle of mass m which involves the scalar field by Eqs. (8.33) and (8.27), or alternatively a conformal factor applied to ordinary spacetime; Eq. (8.36) says there is an 'energy-momentum' tensor which is actually a dynamical object and defines the motion of the particle via the Bianchi identities $\nabla_\beta T^\beta_\alpha = 0$. The last point may stand some elaboration [10]. We note that $\nabla_\beta \left(p^\alpha p^\beta - m^2 g^{\alpha\beta}/2 \right) = 0$ gives $\nabla_\beta p^\beta = 0$, which substituted back gives $\nabla_\beta p^\alpha p^\beta = 0$, the standard 4D equations of motion; and it can be shown by some tedious algebra that the 5D geodesic for the metric (8.27) leads to the same equations. Lastly, we see that the trace of Eq. (8.34) yields $R = -6\varepsilon^2 m^2/h^2$, which agrees with our starting relation (8.19) and again implies that the mass of a particle is related to the scalar curvature of the 4D space it inhabits.

To go further, we need exact solutions. The most convenient way to formulate these is to write the 4D part of the metric as

$$ds^2 = dt^2 - e^{2\lambda} dx^2 - e^{2\mu} dy^2 - dz^2 \quad , \tag{8.37}$$

where $\lambda = \lambda(t, z)$ and $\mu = \mu(t, z)$. This metric can describe a plane wave moving in the z-direction [12, 13]. We expect the associated particle to have momenta

$$p_0 = m\frac{dt}{ds} \equiv E \quad , \quad p_1 = 0 \quad , \quad p_2 = 0 \quad ,$$

$$p_3 = -m\frac{dz}{ds} \equiv -p \quad . \tag{8.38}$$

These particle properties are connected to the wave properties by the field equations (8.34), (8.35). The components of Eq. (8.34), using Eqs. (8.37) and (8.38), read

$$R_{00} = -\partial_0[\partial_0(\lambda + \mu)] - [(\partial_0\lambda)^2 + (\partial_0\mu)^2]$$
$$= -6\varepsilon^2 h^{-2} E^2 \ ,$$

$$R_{03} = -\partial_0[\partial_3(\lambda + \mu)] - (\partial_0\lambda\,\partial_3\lambda + \partial_0\mu\,\partial_3\mu)$$
$$= 6\varepsilon^2 h^{-2} Ep \ ,$$

$$R_{33} = -\partial_3[\partial_3(\lambda + \mu)] - [(\partial_3\lambda)^2 + (\partial_3\mu)^2]$$
$$= -6\varepsilon^2 h^{-2} p^2 \ ,$$

(8.39)

$$R_{11} = e^{2\lambda}[\partial_0(\partial_0\lambda) - \partial_3(\partial_3\lambda) + \partial_0\lambda\,\partial_0(\lambda + \mu) -$$
$$- \partial_3\lambda\,\partial_3(\lambda + \mu)] = 0 \ ,$$

$$R_{22} = e^{2\mu}[\partial_0(\partial_0\mu) - \partial_3(\partial_3\mu) + \partial_0\mu\,\partial_0(\lambda + \mu) -$$
$$- \partial_3\mu\,\partial_3(\lambda + \mu)] = 0 \ ,$$

$$R_{\alpha\beta} = 0 \ \text{(otherwise)} \ .$$

It is obvious when written in this way that there is a solution given by

$$\lambda = -\mu = \sqrt{3}\varepsilon h^{-1}(Et - pz) \ . \tag{8.40}$$

This with Eq. (8.31) gives the action and wave function as

$$I = \int p_\alpha \, dx^\alpha = Et - pz \ , \tag{8.41}$$

$$\Psi = e^{-i(Et - pz)/h} \ . \tag{8.42}$$

The last should satisfy the fifth or Klein-Gordon part of the field equations (8.35), and we find that it does so. The induced or effective 4D energy-momentum tensor by Eqs. (8.36) and (8.38) has the following nonzero components:

$$8\pi T_0^0 = -6\varepsilon^2 h^{-2}(E^2 - m^2/2) \ ,$$

$$8\pi T_3^0 = -8\pi T_0^3 = 6\varepsilon^2 h^{-2} Ep \ ,$$

$$8\pi T_1^1 = 8\pi T_2^2 = 3\varepsilon^2 h^{-2} m^2 \ , \tag{8.43}$$

$$8\pi T_3^3 = 6\varepsilon^2 h^{-2}(p^2 + m^2/2) \ .$$

These obey $8\pi T_\alpha^\alpha = 6\varepsilon^2 h^{-2} m^2$, and we recall that E, p, m are related by $p_\alpha p^\alpha = E^2 - p^2 = m^2$. This constraint means that we have a class of exact solutions that depends on two independent dynamical constants, E and p. [The dimensionless constant ε which we introduced in Eq. (8.33) to connect the scalar field potential and the wave function could in principle be absorbed into E and p, or into h.] We can sum up the 5D wave solution as follows:

$$dS^2 = \Psi^{2\varepsilon} ds^2 - \Psi^{-4\varepsilon} d\ell^2 \ ,$$

$$ds^2 = dt^2 - e^{2\lambda} dx^2 - e^{-2\lambda} dy^2 - dz^2 \ , \tag{8.44}$$

$$\Psi = e^{-i(Et - pz)/h} \ , \quad \lambda = \sqrt{3}\varepsilon h^{-1}(Et - pz) \ .$$

We have examined algebraic properties of this class of solutions using a software package. Apart from confirming $R_{AB} = 0$, we find $R_{ABCD} \neq 0$ so these solutions are not 4D curved and 5D flat, like those considered in Sections 3.2 and 4.5. (For example, for the above solutions we find that $R_{1414} = -2\varepsilon^2(3E^2 + p^2)\Psi^{-4\varepsilon} \neq 0$ and also $K \equiv R_{ABCD}R^{ABCD} \neq 0$.) However, while these solutions are in general curved, they have $\sqrt{-g} = 1$ for the determinant of the 4D metric and so are special algebraically.

There is, however, a physical characteristic of the above solutions which is expected to have widespread application. Namely, the fact that the effective energy-momentum tensor is not gravitational but rather quantum in nature. To balance Einstein's equations, it is necessary to have a quantity on the right-hand side which has the physical dimensions (units) of (length)$^{-2}$ and reflects the nature of the source. In gravitational problems, this is typically the density of the fluid in the form $G\rho/c^2$. For quantum problems, we see that the source typically involves the particle mass in the combination $(mc/h)^2$. This may seem odd, but as far back as

1935 Dirac proposed a neat way of categorizing the mass and other properties of a particle by using a 5D version of 4D de Sitter space [14]. Also, in more recent times Matute suggested that the classical and quantum aspects of matter represent a kind of duality [15]. If the existence of two kinds of energy-momenta tensors, typified by G and h, causes unease, then wave mechanics may be viewed as a kind of bridge: a discrete particle has de Broglie waves which formally spread throughout ordinary 3D space, so the localized mass becomes a kind of global fluid via the concept of wave-particle duality.

8.5. Wave Mechanics in 5D

In this section and the rest, we revert to the (pure) canonical metric of Chapter 7, and use its regular and shifted forms to construct models for de Broglie waves and quantized particles. The use of this metric implies that we are studying the physics of the vacuum, with its characteristic equation of state where the pressure and density sum to zero. These physical quantities are measured by the cosmological constant Λ, and we recall that the magnitude of this is subject to controversy (see Chapters 1 and 7): while Λ is small and positive on cosmological scales, it could be large and negative on sub-nuclear scales. The aim of the present section is to see if the peculiar properties of 4D wave mechanics, such as superluminal velocities, can be better explained by a 5D approach.

The theoretical basis of wave mechanics was stated by de Broglie in 1923. He postulated that any massive particle has associated with it a wave, analogous to the photon's electromagnetic wave, in a way that respects the Lorentz transformations and Planck's law. The direction of motion of the particle is the same as the normal to wave-front. Let the mass and velocity of the particle be m, v and the frequency and velocity of the wave be f, w. Then in terms of the 4-vectors for the particle and the wave, there is a match of the magnitude of the quantities concerned, via de Broglie's equation: $m(v, 1) = hf(1/w, 1/c^2)$. Equating components in de Broglie's equation gives $mv = hf/w$ and $m = hf/c^2$. The second of these relations is simply a statement of the equality of the energy of the particle and the energy of the wave: $E = mc^2 = hf$. The first relation, when divided by the second, gives $vw = c^2$.

Since a normal particle has $v < c$, de Broglie's velocity equation necessarily implies some kind of wave with a phase velocity $w > c$. This makes many workers uncomfortable.

An argument has, however, been made by Rindler [16] for the compatibility of superluminal velocities and the Lorentz transformations. Consider the usual setup in which two inertial frames F and F' are in relative motion along the x-axis at velocity $v = dx/dt$. The relevant Lorentz transformation relates the times in the two frames via $t' = (1 - v^2/c^2)^{-1/2}(t - vx/c^2)$. Then a wave which is emitted by a particle in F' at $t' = 0$ has (in F) the equation of motion given by $t = vx/c^2$. This gives the speed of the wave as $w \equiv x/t = c^2/v$ (as observed in F). That is, $vw = c^2$ as above.

This argument is technically correct, but the implication of superluminal propagation is still difficult to accept, as are many aspects of quantum mechanics [17]. However, waves of this type arise inevitably in any 5D theory of the kind used in unification, if the extra coordinate is timelike. Consider flat 5D space with line element $dS^2 = c^2 dt^2 - (dx^2 + dy^2 + dz^2) + d\ell^2$. For the 5D null path $dS^2 = 0$, the apparent velocity along the x-axis has magnitude $dx/dt = [c^2 + (d\ell/dt)^2]^{1/2}$, and can exceed c. For more realistic situations, in which there is a significant vacuum measured by a cosmological constant $\Lambda < 0$, superluminal waves are a natural consequence. From the viewpoint of 5D field theory, waves of de Broglie type have to be considered real.

That de Broglie waves are real is in fact shown by modern experiments [18]. Recent refinements of the classic double-slit experiment, and other setups, have confirmed the wave nature of matter for relatively large molecules (e.g., carbon-60). For a composite object, the de Broglie wavelength depends on its constituents, and there is a regular transition of the wave properties from massive particles to entangled photons. Yet despite extensive experimental work, the origin and nature of these waves remains mysterious [19].

The prime reason for this is that the vacuum with $\Lambda < 0$ has an energy density $\rho_v < 0$, and such a medium is unlike any form of ordinary matter. All waves in ordinary matter have velocities given by an expression of the sort $w = \sqrt{K/\rho}$, where K is a constant that depends on microscopic physics (the bulk modulus for fluids and the shear modulus for solids). This is discussed in standard texts, a few of which point out that a value $K < 0$ would imply that an increase in the ambient pressure causes an increase in the volume of an element of the material, or equivalently a decrease in the density. This behaviour is the opposite of that observed in ordinary matter. However, it is exactly the behaviour consistent with the equation of state $p_v = -\rho_v c^2$ of the vacuum when the pressure is positive and the density is negative.

The canonical metric is the natural format for studying the vacuum in 5D, since we recall that it gives back all Einstein spaces with Λ in 4D (Chapter 7). There are many exact solutions of the 5D field equations $R_{AB} = 0$ of canonical form, including ones of de Sitter type. These have been applied to the solar system and inflationary cosmology (Sections 7.5, 7.6), where the assumption is that $\Lambda > 0$. The inflationary aspects of the early universe, in particular, have been intensively studied by Bellini and coworkers [20], who concluded that the 5D canonical metric is acceptable as a basis for 4D inflation. However, it is also possible to have canonical solutions which possess $\Lambda < 0$ and describe local physics. We will employ such a solution to construct a model of a quantized particle in the next section. Here, though, the focus is on waves; and we revisit the solution of this type considered in Section 4.6 for astrophysics, and inquire if it can be applied to de Broglie waves.

Consider, then, the solution to $R_{AB} = 0$ given by

$$dS^2 = \frac{\ell^2}{L^2}\left\{dt^2 - \exp\left[\pm\frac{2i}{L}(t + \alpha x)\right]dx^2 - \exp\left[\pm\frac{2i}{L}(t + \beta y)\right]dy^2 - \exp\left[\pm\frac{2i}{L}(t + \gamma z)\right]dz^2\right\} + d\ell^2 \quad . \quad (8.45)$$

This describes a wave propagating through ordinary 3D space, where the frequency $f = 1/L$ is fixed by the solution. The wave numbers $k_x = \alpha/L$, $k_y = \beta/L$, $k_z = \gamma/L$ along the x, y, z axes are written in terms of dimensionless constants α, β, γ which are arbitrary. The phase velocity of the wave along (say) the x-axis is c/α, and is also arbitrary. By evaluating the Einstein tensor from the 4D part of Eq. (8.45), it may be shown by some algebra that the equations of general relativity are satisfied, with a cosmological constant $\Lambda = -3/L^2$. Accordingly, Eq. (8.45) describes waves travelling in a classical vacuum with positive pressure.

To investigate further, consider the element of proper distance in Eq. (8.45) along the x-axis given by $d\bar{x} = \exp[i(ft + k_x x)]dx$. The frequency is $f = c/L$. By Planck's law, the energy of the wave is equivalent to the mass m of the associated particle, $E = hf = hc/L = mc^2$, so $L = h/mc$. That is, the size of the 4D potential well in Eq. (8.45) equals the Compton wavelength of the particle. The momentum of the particle $p_x = mv_x$ is inversely proportional to the wavelength, so the wave number k_x is directly proportional to p_x, and can be written in the correct dimensional form as $k_x = (mc/h)(v_x/c) = v_x/cL$. Recalling that the frequency is $f = c/L$, the phase velocity of the wave along the x-axis is $w_x = f/k_x = (c/L)(cL/v_x) = c^2/v_x$. Thus along each axis, the velocities of the particle and the wave are connected by de Broglie's relation $vw = c^2$.

The question arises of whether a de Broglie wave is longitudinal (like a sound wave) or transverse (like a light wave). To answer this, we refer back to Section 7.4, where we discussed the fifth force which attends any ℓ-dependent 5D metric. This force acts in spacetime, parallel to the 4-velocity u^μ, and in general form is given by

$$P^\mu = \left(-\frac{1}{2}\frac{\partial \gamma_{\alpha\beta}}{\partial \ell}u^\alpha u^\beta\right)\frac{d\ell}{ds}u^\mu \ . \tag{8.46}$$

Here, the term in parentheses represents the coupling between 4D and the extra dimension, and $\gamma_{\alpha\beta} = \gamma_{\alpha\beta}(x^\gamma, \ell)$ is the 4D metric tensor in unfactorized form. We noted before that the force (per unit mass) or acceleration (8.46) is gauge-dependent. That is, it depends on the choice of coordinates. For the pure canonical metric, there is a special choice of coordinates that renders P^μ equal to zero. However, for the present application we expect P^μ to be the force responsible for the perturbations in the vacuum which we call de Broglie waves, so it is finite. The magnitude is determined by putting $\gamma_{\alpha\beta} = (\ell/L)^2 g_{\alpha\beta}(x^\gamma)$ in Eq. (8.46), which gives $P^\mu = -(1/\ell)(d\ell/ds)u^\mu$. The minus sign here represents a restoring force towards the local centre of motion. The form of $d\ell/ds$ in the last expression is obtained by putting the null-path condition $dS^2 = 0$ in Eq. (8.45), which gives $\ell = \ell_* \exp(\pm is/L)$. In this, ℓ_* is the amplitude of the wave and L is its wavelength, which we determined before to be $\ell =$

h/mc. We find that $P^\mu = \pm iu^\mu/L$, where the 4-velocity is complex. The precise form of u^μ can be obtained by equating P^μ to the local acceleration du^μ/ds, and integrating. The resulting expression can be integrated again to obtain, finally, the form of $x^\mu = x^\mu(s)$. It is a sine wave. This result might have been expected, since sinusoidal motion is common for metrics of canonical type. However, some interpretation is required in the present case. Thus, while ℓ oscillates out of the s-surface we call spacetime, the fifth force P^μ acts *in* the surface of spacetime. It causes a point on the x-axis (say) to oscillate about a location on the overall path, which is fixed by the apparatus in a practical setup like that of the double-slit experiment. The net consequence of these considerations is that we expect the de Broglie wave observed in the laboratory to be longitudinal in nature.

Another question which arises in connection with the wave solution (8.45) is whether it fits into the general scheme of wave mechanics, and in particular how solutions of that type relate to the Klein-Gordon equation. We touched on the latter relation in Sections 8.1 and 8.4, and a brief account may be instructive here. Thus given a solution with a well-defined wavelength L, it is always possible to define a dimensionless action, $I \equiv \int ds/L = \int (h/mc)^{-1} ds$. From this, a wave function may if so desired be formed, $\psi = \exp(iI)$. The first derivative of this gives $p_\alpha = (h/i\psi)\,\partial\psi/\partial x^\alpha$, where the 4-momenta are defined as usual ($p^\alpha \equiv mc\,u^\alpha = mc\,dx^\alpha/ds$). The second derivative, taken covariantly if the spacetime is curved, splits into a real part and an imaginary part. One of these gives $\nabla_\beta p^\beta = 0$, the standard conservation law for the momenta. The other part gives $\Box\psi + (c/h)^2 m^2 \psi = 0$, where $\Box\psi \equiv g^{\alpha\beta}\nabla_\beta(\partial_\alpha\psi)$. This last relation is the Klein-Gordon equation.

Other implications of the wave solution (8.45) have been discussed elsewhere [21]. That solution was found by hand but may be verified by computer and is somewhat unique in being algebraically simple and having a relatively straightforward physical interpretation. However, we can reasonably expect the vacuum to be occupied by other types of wave, with forms depending on the local physics. In this regard, it should be recalled that in 5D theory the cosmological 'constant' is generally a function of ℓ, and via $\ell = \ell(s)$ for the path and $s = s(x^\gamma)$, also a function of the spacetime coordinates (see Chapter 7). For the pure canonical metric, the 5D field equations give in the limit of no ordinary matter $\Lambda = \pm 3/\ell^2$, which in the hypersurface $\ell = L$ where spacetime is recovered and Einstein's equations hold gives the familiar $\Lambda = \pm 3/L^2$. Clearly, Λ may vary from one ℓ-hypersurface to another, offering in 5D a potential resolution of the so-called cosmological-'constant' problem. Also, as we saw in Section 8.4, it is possible that quantum theory employs a different field-matter coupling than classical theory. The latter defines the properties of the vacuum via $p_v = -\rho_v = \pm \Lambda c^2/8\pi G$. The equation of state here is typical of the Einstein vacuum, and should be preserved even when the problem concerned is not gravitational but quantum in nature. However, in that case we expect that the defining parameter should not be G but rather h. The relevant expression for the magnitude of vacuum density, up to a dimensionless factor, is then $|\rho_v| = h\Lambda^2/8\pi c$. For the metric (8.45), $\Lambda = -3/L^2$ where $L = h/mc$, so the vacuum density is

$$|\rho_v| = \frac{h\Lambda^2}{8\pi c} = \left(\frac{9}{8\pi}\right)\frac{c^3 m^4}{h^3} = \frac{3}{2}\frac{m}{(4\pi\lambda_C^3/3)} \ . \tag{8.47}$$

Here $\lambda_C = h/mc$ is the Compton wavelength. The implication of this expression is clear: the effective density experienced by the vacuum wave is the mass of the particle divided by the volume of its Compton sphere.

This is logical, and together with the other results found above leads us to conclude that the metric (8.45) is a good 5D solution for a 4D de Broglie wave. The puzzling properties of the latter in 4D are seen to be due to the truncation of a well-behaved solution in 5D.

Wave-particle duality, however, remains problematic. We will meet this again in the following section, where we will apply a shift $\ell \to (\ell - \ell_0)$ to the canonical metric, which then takes the form $dS^2 = [(\ell - \ell_0)/L]^2 ds^2 \pm d\ell^2$. Here ds^2 is the 4D metric of any vacuum Einstein space, and the sign choice indicates a timelike ($g_{44} > 0$) or spacelike ($g_{44} < 0$) extra dimension. Depending on this choice, the path $\ell(s)$ for a null geodesic with $dS^2 = 0$ is either oscillatory or monotonic. With (i) to indicate the two cases, they are given by $\ell = \ell + \ell_* \exp[\pm(i)s/L]$. At first sight, it may appear that the oscillatory and monotonic modes can be simply identified with waves and particles, and that the two together offer an explanation of wave-particle duality. However, this meets a problem which is of long standing in the foundations of quantum mechanics: Do the wave and the particle exist as separate things, or are they aspects of the *same* thing? The exact solution (8.45) describes only a wave, and as such has $g_{44} > 1$ and $\Lambda < 0$ in the canonical metric. Conversely, the corresponding particle has $g_{44} < 1$ and $\Lambda > 0$. These solutions are separate, and each implies significant spacetime curvature. The curvature contributions would only cancel if both modes were to occur simultaneously, as the result of an horizon in the fifth dimension or a change in the mode of the complex scalar field associated with it. We see that the knotty problem of wave-particle duality needs a bit more thought.

8.6. A Quantized Particle

In the preceding section, it was shown that the properties of de Broglie waves follow from a certain form of the canonical metric. Those waves propagate through spacetime, or are free. In the present section, attention is given to waves which are confined to a region of spacetime, or are trapped.

It is important to realize that there is a difference between a model for a particle and a model that has quantization. Objects with particle-like properties can be obtained even in 4D general relativity, as mentioned before. And in 5D, the solitons of Chapter 5 might with modifications be regarded as particles. There are also other solutions which could serve this purpose [22]. It should be noted in this regard that the existence of the Klein-Gordon equation does not by itself imply quantization. For the (pure) canonical metric, it may be shown by some algebra that the extra component of the 5D geodesic equation is equivalent to the Klein-Gordon equation [23]. It may likewise be shown that when the prefactor of the canonical metric is treated as a 4D conformal factor, the 4D curvatures with and without the factor are related by an equation involving the mass which agrees with the argument of Section 8.4. However, none of this work results in quantization. The reason, it turns out, is that quantization entails waves which are trapped in the geometry by a negative divergence in the energy density of the vacuum.

The appropriate metric, by Chapter 7, is the canonical one with a shift $\ell \to (\ell - \ell_0)$ in the extra coordinate, where ℓ_0 is a constant length. The line element and its attendant cosmological 'constant' are

$$dS^2 = \left(\frac{\ell - \ell_0}{L}\right)^2 ds^2 \pm d\ell^2 \;,$$

$$ds^2 = g_{\alpha\beta}(x^\gamma)dx^\alpha dx^\beta \;, \tag{8.48}$$

$$\Lambda = \mp \frac{3}{L^2}\left(\frac{\ell}{\ell - \ell_0}\right)^2 \;.$$

In this, the null path $dS^2 = 0$ gives as before the motions $\ell = \ell(s)$:

$$\ell = \ell_0 + \ell_* \exp(\pm s/L) \quad (\ell \text{ spacelike}) \tag{8.49a}$$

$$\ell = \ell_0 + \ell_* \exp(\pm is/L) \quad (\ell \text{ timelike}) \;. \tag{8.49b}$$

Here ℓ_* is a constant and the sign choice merely reflects the reversibility of the motion in the extra dimension. These motions apply to a wide variety of vacuum solutions to the Einstein equations. But the simplest such solution is the local one of de Sitter, which exists with both signs of Λ. This will be taken below as the default 4D kernel for the 5D metric. With present terminology, the 4D Gaussian curvature for de Sitter is $K = -1/L^2$ for $\Lambda > 0$ and $K = +1/L^2$ for $\Lambda < 0$.

Figure 8.1 is a plot of the function $\Lambda(\ell)$ as given by Eqs. (8.48), though not all parts of that plot may be accessible to a real particle or wave with the motions $\ell(s)$ of Eqs. (8.49). Some features are striking. For $\Lambda > 0$, this parameter is divergent for $\ell = \ell_0$ ($s = 0$), and decays with $\Lambda \to +3/L^2$ for $\ell \to \infty$ ($s \to \infty$), tending to the value typical of the unshifted metric. For $\Lambda < 0$, a wider variety of behavior is possible. The constants ℓ_0 and ℓ_* are arbitrary in Eqs. (8.48) and (8.49), but might be constrained by topology in practice. Basically the motion is oscillatory about ℓ_0, with $(\ell_0 - \ell_*) \le \ell \le (\ell_0 + \ell_*)$. Formally, Λ is divergent at $\ell = \ell_0$, but the wave can traverse this. Effectively, the wave is trapped in the 'groove' formed by the negative divergence of Λ. The *average* value of Λ can be

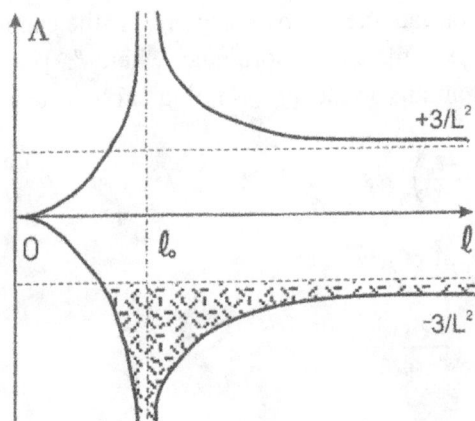

Figure 8.1: A schematic plot of the cosmological 'constant' Λ as a function of the extra coordinate ℓ, according to Eq. (8.48) of the text. The stippled region is a trap for the wave defined by Eq. (8.49b).

calculated by inserting the root-mean-square value of Eq. (8.49b) into Eq. (8.48). The result is

$$\overline{\Lambda} = -\frac{3}{L^2} \quad , \quad \frac{\ell_0}{\ell_*} \ll 1 \quad ,$$

$$\overline{\Lambda} = -\frac{6}{L^2}\left(\frac{\ell_0}{\ell_*}\right)^2 \quad , \quad \frac{\ell_0}{\ell_*} \gg 1 \quad .$$

(8.50)

The second of these applies to a wave whose amplitude is small compared to the shift, a situation which is physically plausible. It is seen that the value of $\overline{\Lambda}$ is boosted above what might have been expected from the L in the metric, by a factor which depends on the ratio of the shift to the amplitude. Physically, this is another expression of the existence of the divergence at $\ell = \ell_0$, and the fact that a small-amplitude wave is preferentially located there. Given that the shifted pure-canonical metric (8.48) allows of both positive and negative values of the cosmological 'constant', it is interesting to consider a model universe consisting of a background with an open topology, studded with small isolated inclusions with closed topologies. The preceding results imply that the background

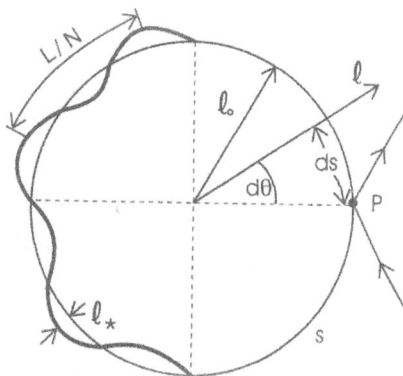

Figure 8.2: A schematic representation of the wave in the extra coordinate ℓ as it propagates around the closed surface of spacetime s, according to Eq. (8.49b). In the latter the fundamental mode is treated, whereas in the figure a harmonic with shorter wavelength is shown. The point P marks the impact of a scattered particle.

has Λ positive and relatively small, while the inclusions have Λ negative and of large magnitude.

It is straightforward to employ the preceding results to construct a toy model of a particle with quantization. In canonical space, the extra coordinate $x^4 = \ell$ is orthogonal to the 4-space with coordinates x^γ, so Eq. (8.49b) represents a wave which oscillates around spacetime. Since the motion involves $\exp(\pm is/L)$, we identify L with the Compton wavelength h/mc of the associated particle, as before. We have $\Lambda < 0$, so the topology is closed, and we take it to be circular as shown in Figure 8.2. Referring to this, the focus of the ℓ-wave is a circle of radius ℓ_0 and the fundamental mode has wavelength L (though overtones are presumably allowed with wavelengths L/N where N is an integer). Clearly, one wave has to fit into the circumference $2\pi\ell_0$, so $2\pi\ell_0 = L = h/mc$ and $\ell_0 = \hbar/mc$. This length fixes the angle $d\theta = ds/\ell_0$ around the orbit from a point P which marks the impact of a scattered particle. Substituting the above expression for ℓ_0 gives the angle as $d\theta = mcds/\hbar$. For n revolutions, one finds:

$$\int_0^{2\pi n} d\theta = \int \frac{mcds}{\hbar} \; ,$$

$$(8.51)$$

$$mcs = nh \quad .$$

This is, of course, the standard rule for quantization.

The model outlined above is simple but effective. The preceding account follows Wesson [24], who has discussed the implications of the model and ways to generalize it. Basically, the model consists of a globule of vacuum-dominated spacetime. This is embedded in a 5D canonical space with a shift whose magnitude determines the size of the globule. The latter is closed in topology, with a groove where the (negative) vacuum energy as measured by the cosmological 'constant' diverges. A wave is trapped in this groove, with a wavelength which determines the mass of the associated particle. The requirement that the trapped wave fit around the globule results in the standard rule for quantization.

8.7. Conclusion

In 5D, spacetime may be viewed as a surface that goes up or down depending on the local value of the energy density of the vacuum. The latter is measured by the cosmological 'constant', which depends in general on the coordinate at right angles to spacetime, so we can talk of a Λ-landscape. This landscape is expected to be smooth and have positive elevation over cosmological distances; but it can show sharp dips at small distances, associated with particles.

We would be able to understand the relationship between physics on the large and small scales if we had a complete theory that unifies gravitation with quantum phenomena. In the absence of this, it makes sense to look at a theory which occupies the middle ground. To this end, we have looked in this chapter at what might loosely be called wave mechanics à la 5D general relativity. It features Planck's constant, but the format is an extension of Einstein's theory.

The belief that there is indeed some connection between physics on the large and small scales is embodied by Mach's Principle (Section 8.2). Mach held the opinion that the rest mass of a local particle ought to be organically connected to other matter, even very remote matter. This implies that particle mass ought to be determined by some kind of global field equation. Formally, the Klein-Gordon equation fulfills this requirement, though it is not usually regarded in this manner.

Another alternative is provided by STM theory, in which there is a special set of coordinates – termed canonical – where the extra coordinate plays the role of particle mass. (In gravitational problems the relevant length is the Schwarzschild radius and in quantum problems it is the Compton wavelength.) There are actually many properties of particles which can be connected to 5D geometry, especially via the vacuum (Section 8.3). If Mach's principle holds, particularly in conjunction with wave mechanics, it might be expected that the rest mass of a particle is related to the scalar curvature of the space it inhabits (Section 8.4). The latter scheme suggests that the effective energy-momenta tensor involves a coupling which depends on Planck's constant rather than the gravitational constant. The same conclusion follows from examining wave mechanics in 5D (Section 8.5). This reveals that the puzzling properties of the de Broglie waves in 4D – such as their superluminal speeds – may arise from truncating a solution which is well-behaved in 5D. Quantization turns out to be an essentially 5D phenomenon (Section 8.6). There is undoubtedly a divergence in the energy density of the vacuum for the shifted-canonical metric, and this can lead to trapped waves. A model for a closed region of vacuum-dominated spacetime leads to the standard quantization rule. This model is admittedly simple (it has the level of complexity of the Bohr model for the hydrogen atom). But like the other things examined in this chapter, it can be developed.

It is only recently that we came to realize that the universe on the large scale is dominated by the vacuum. From a consideration of the behaviour of particles and waves in relation to 5D, we infer that the world on the small scale may also by dominated by the vacuum.

References

[1] Mach, E., *The Science of Mechanics* (Open Court, La Salle, 1960).

[2] Wesson, P.S., *Cosmology and Geophysics* (Hilger /Oxford Un. Press, Bristol/New York, 1978); Barbour, J.B. and Pfister, H., *Mach's Principle: from Newton's Bucket to Quantum Gravity* (Birkhäuser, Basel, 1995). For more recent discussions see Overduin, J.M. and Fahr, H.-J., *Naturwissenschaften* **88**, 491 (2001); Halpern, P. and Wesson, P., *Brave New Universe* (J. Henry Press, Washington, 2006).

[3] Wesson, P.S., *Weaving the Universe: Is Modern Cosmology Discovered or Invented?* (World Scientific, Singapore, 2011).

[4] Mashhoon, B. and Wesson, P.S., *Ann. Phys.* (Berlin) **524**, 63 (2012); Barbour, J.B., *Ann. Phys.* (Berlin) **524**, A39 (2012); Mashhoon, B. and Wesson, P.S., *Ann. Phys.* (Berlin) **524**, A44 (2012).

[5] Mashhoon, B., Liu, H. and Wesson, P.S., *Phys. Lett.* **B331**, 305 (1994).

[6] Wesson, P.S., *Gen. Rel. Grav.* **35**, 111 (2003).

[7] Wesson, P.S., *Mod. Phys. Lett.* **A19**, 1995 (2004).

[8] Wesson, P.S., *Class. Quant. Grav.* **19**, 2025 (2002).

[9] Wald, R.M., *General Relativity* (Un. Chicago Press, Chicago, 1984), p. 446.

[10] Liu, H. and Wesson, P.S., *Int. J. Mod. Phys.* **D7**, 737 (1998).

[11] Mandl, F. and Shaw, G., *Quantum Field Theory* (Wiley, New York, 1993), p. 29.

[12] Liu, H. and Wanzhong, S., *Gen. Rel. Grav.* **20**, 407 (1988).

[13] Kramer, D., Stephani, H., MacCallum, M. and Herlt, E., *Exact Solutions of Einstein's Field Equations* (Cambridge Un. Press, Cambridge, 1980).

[14] Dirac, P.A.M., *Ann. Math.* **36**, 657 (1935); **37**, 429 (1936).

[15] Matute, E.A., *Class. Quant. Grav.* **14**, 2771 (1997).

[16] Rindler, W., *Essential Relativity,* 2nd. ed. (Springer, New York, 1977), p. 91.

[17] Bell, J.S., *Speakable and Unspeakable in Quantum Mechanics*, 2nd. ed. (Cambridge Un. Press, Cambridge, 2004); Halpern, P., *The Pursuit of Destiny: A History of Prediction* (Perseus Press, Cambridge, 2000).

[18] Kocsis, S., Braverman, B., Roberts, S., Stevens, M.J., Mirin, R.P., Shalm, L.K. and Steinberg, A.M., *Science* **332**, 1170 (2011); Edamatsu, K., Shimizu, R. and Itoh, T., *Phys. Rev. Lett.* **89**, 213601 (2002); Colella, R., Overhauser, A.W. and Werner, S.A., *Phys. Rev. Lett.* **34**, 1472 (1975); Rauch, H. and Werner, S.A., *Neutron Interferometry* (Clarendon Press, Oxford, 2000).

[19] Wesson, P.S., *Mod. Phys. Lett.* **A29**, 1450168 (2014); Wesson, P.S. and Overduin, J.M., *Phys. Lett.* B750, **302** (2015); Wesson, P.S., *Ann. Phys.* (Berlin) **528**, 307.

[20] Bellini, M., *Nucl. Phys.* **B660**, 389 (2003); *Phys. Lett.* **B632**, 6 (2011).

[21] Wesson, P.S., *Phys. Lett.* **B722**, 1 (2013).

[22] Israelit, M., *Found Phys.* **37**, 1628 (2007); *Gen. Rel. Grav.* **40**, 2469 (2008).

[23] Wesson, P.S., *Phys. Lett.* **B701**, 379 (2011).

[24] Wesson, P.S., *Phys. Lett.* **B706**, 1 (2011).

Chapter 9

The Fifth Dimension

9.1. Unification and Geometrization

Space-Time-Matter theory is based on the proposition that four-dimensional (4D) matter and energy are manifestations of the shape of a higher-dimensional world that is empty. The idea itself is an old one, perhaps first clearly expressed by the mathematician William Kingdon Clifford. Deeply impressed by Bernhard Riemann's discovery of non-Euclidean geometry, Clifford addressed the Cambridge Philosophical Society in 1870 as follows: "I hold ... that this variation of the curvature of space is what really happens in that phenomenon which we call the motion of matter ... *and that in the physical world nothing else takes place but this variation*" ([1], emphasis ours).

As described in Chapter 1, Einstein's 4D theory of general relativity is a partial realization of Clifford's vision, in that the geometry of spacetime (the left-hand side of the field equations) is equated mathematically with its matter-energy content (the right-hand side). But the two terms remain logically distinct. Einstein himself was deeply unsatisfied with this dualistic aspect of his theory, and, as is well known, devoted his last decades to the quest for a more general unified theory.

It is widely agreed that he failed. Confusion lingers, however, about what exactly would have constituted success in Einstein's eyes. For many physicists, the geometrization of matter is the crowning glory of general relativity, and presumably also of any more general theory that might someday subsume it. Misner, Thorne and Wheeler, for example, dedicated their classic text to "Einstein's inspiring vision of geometry as the machinery of physics" [2], and many of the best modern books use similar language [3]. Einstein himself referred to geometry as "the oldest branch of physics" [4] and famously compared the field equations of his theory "to a building, one wing of which is made of fine marble (left part of the equation), but the other wing of which is built of low-grade wood (right side of equation)" [5]. But the philosopher Hans Reichenbach, who corresponded regularly with Einstein on the subject, countered that general relativity does not so much geometrize physics as it does *physicalize geometry* [6]. And Peter Bergmann, Einstein's closest partner in the search for a unified theory, insisted that his mentor "stated repeatedly that he did not consider geometrization of physics a foremost, or even a meaningful objective ... What really counts is not a geometric formulation, or picturization, but a real fusing of the mathematical structures intended to represent physical fields" [7].

Recent work by historians of science leaves little doubt that Bergmann was right [8]. What Einstein sought above all was unification of the gravitational (and other) fields with their sources, not geometrization *per se*. Two years before his death, he wrote to a young protégé, John Moffat: "I feel compelled to consider logical simplicity alone as a guide ... Essential to this is the view that the Lagrange function should not be composed of logically independent parts" [9]. He echoed this theme in an Appendix to a posthumous edition of *The Meaning of Relativity*: "The present theory of relativity is based on a division of physical reality into a metric field (gravitation) on the one hand, and into an electromagnetic field and matter on the other hand. In reality space will probably be of a uniform character and the present theory be valid only as a limiting case" [10].

Space-Time-Matter theory is a logically minimal extension of general relativity that realizes Einstein's dream in both senses. The metric structure of the 4D theory is inherited and merely extended to one additional coordinate. Fields are unified physically with their sources by deriving the 4D field equations with matter from the 5D Ricci equations in vacuum. That this can always be done is guaranteed by Campbell's embedding theorem (Chapter 2). It is intriguing that this theorem appeared in 1926, the same year as Klein's compactification of Kaluza's fifth dimension. Compactified Kaluza-Klein theory went on to become the basis for most work on higher-dimensional unification, while Campbell's theorem languished in obscurity. Einstein, as we know, worked on his own 5D theories with Jakob Grommer in the 1920s [11] and Peter Bergmann and Valery Bargmann in the 1930s [12]. Nowhere did he express the nature of his quest for unification more clearly than in his paper with Grommer [13]:

> "We presently have in mind as ultimate goal *a pure field theory, in which the field variables produce the field of 'empty space' as well as the ... elementary particles that constitute 'matter'.*"

With Bergmann, Einstein followed an approach reminiscent of Klein's, stressing that *"we ascribe physical reality to the fifth dimension"* ([12], emphasis ours). Yet he ultimately abandoned higher dimensions as a basis for unification, possibly because he believed that they could not lead to nonsingular models for elementary particles [13], or because he was reluctant to introduce a new long-range scalar field into physics [14]. What might Einstein have done had he known of Campbell's embedding theorem? He might, of course, have dismissed it much as he initially dismissed Minkowski's 4D formulation of special relativity: as "superfluous learnedness" [15].

Space-Time-Matter theory is so far incomplete, in that it gives an account only of gravity, electromagnetism, and a scalar field. The induced 4D matter sources are describable mostly in terms of macroscopic properties such as energy density, pressure, and bulk equation of state (Chapters 3-5). The nature and role of the scalar field are not yet clear (Chapters 6, 8). A comprehensive account of fermionic matter and elementary particle spin remains elusive (Chapters 7, 8). This latter issue has proved challenging for other geometrical unified theories, including those based on geons [16], kinks [17], torsion [18, 19], Clifford algebras [20] and sigma models [21]. Further progress in all three areas may depend on the discovery of new exact solutions of the field equations in higher dimensions with appropriate properties on 4D hypersurfaces. There remains rich scope for further inquiry on the theoretical side.

On the experimental side, the theory is consistent with observation, but as yet lacks a "smoking gun" that could distinguish it unambiguously from 4D GR. A degree of ambiguity necessarily arises because the same 4D physics can be embedded differently in different solutions of the 5D field equations. Thus the classical tests of GR, for example, were originally applied to a 5D generalization of the Schwarzschild metric (the soliton metric), leading to very strong constraints on a primary dimensionless free parameter of the theory, of order parts in 10^4 [22, 23] or even 10^8 or more [24, 25]. But later work has established that the 5D canonical metric provides a more appropriate basis for comparison with 4D physics (Chapter 7), and constraints in this case are weaker [26, 27]. However, it is important not to lose sight of the fact that *a 5D-covariant theory must necessarily violate the 4D conservation laws at some level, and must therefore be testable in principle.* The question then boils down to why the fifth dimension has not been seen to date, and indeed whether it is possible to see it at all in practice.

The answer to this question may lie, not in the new mathematical degree of freedom that the fifth dimension represents, but in its physical interpretation. If so, then the right place to begin looking is not with Theodor Kaluza in 1921, but rather with Hermann Minkowski in 1908.

9.2. Beyond Space and Time

Minkowski, as is well known, introduced time as the fourth dimension at a meeting in Cologne with the words [28]:

> "The views of space and time which I wish to lay before you have sprung from the soil of experimental physics, and therein lies their strength. They are radical. Henceforth space by itself, and time by itself, are doomed to fade away into mere shadows, and only a kind of union of the two will preserve an independent reality."

His proposal was indeed radical. By uniting space and time in a common metrical framework, Minkowski shattered the unconscious Pythagorean prejudice that geometry applies only to *length-like quantities*. But if this view sprang from the soil of experimental physics, why did its discovery take so long?

In fact, there were intriguing precursors to Minkowski's idea, going back to as early as the 18th century when Joseph-Louis Lagrange observed in the *Theory of Analytical Functions* that with time as a fourth coordinate, "one can regard mechanics as four-dimensional geometry." Others who made similar (if sometimes vague) suggestions included the philosopher Arthur Schopenhauer in 1818, the physicist and mathematician William Rowan Hamilton in 1843, the poet Edgar Allan Poe in 1848, and the composer Richard Wagner in 1877 (see Refs. [29, 30] for reviews). Perhaps the most explicit of these was "S.," an anonymous contributor to *Nature* in 1885, who observed that "...there is a new three-dimensional space for each successive instant of time; and, by picturing to ourselves the aggregate formed by the successive positions in time-space of a given solid during a given time, we shall get the idea of a four-dimensional solid..." [31]. "S." is thought to have been the mathematician James Joseph Sylvester [32].

Minkowski's geometrization of time via $x^0 = ct$ was, of course, motivated by Einstein's 1905 reconciliation of Newtonian mechanics with Maxwellian electromagnetism in the form of special relativity (SR). What was truly radical, and what had escaped his predecessors, was the *minus sign* in the expression for distance between nearby events in space and time:

$$ds^2 = -c^2 dt^2 + |d\vec{x}|^2 \ , \tag{9.1}$$

The reason why it took so long to recognize that the Pythagorean theorem is actually a 4D relation, not a 3D one, is that the dimension-transposing constant "c" that converts time into units of distance in Eq. (9.1) is much larger than characteristic speeds in everyday life (or equivalently, because the trajectories of massive bodies in spacetime lie close to the t-axis). The effects of the new coordinate show up in 4D SR mainly through the Lorentz factor arising from Eq. (9.1):

$$\gamma_4 \equiv \frac{cdt}{d\tau} = \frac{1}{\sqrt{1 - (|d\vec{x}|/cdt)^2}} \ , \tag{9.2}$$

where $d\tau^2 \equiv -ds^2$. If $|d\vec{x}|/dt \ll c$, as is true in most of human experience, then $\gamma_4 \approx 1$ and spacetime looks like space.

Kaluza may have been inspired by Minkowski's example when he conceived the idea of further unifying Maxwellian electromagnetism with Einsteinian gravity by means of a fifth spatial dimension, $x^5 = \ell$. Gunnar Nordström had the same idea some years earlier, but he used a pre-general relativistic scalar theory of gravity and his work has been largely forgotten. To explain why physics appeared to be independent of x^5, Kaluza was obliged to impose his "cylinder condition," thus hobbling the new theory from the outset. Klein subsequently found a physical motivation for this condition in the assumption of compactification. Extended further to additional spatial dimensions, compactified Kaluza-Klein theory then became the basis for most work in unified-field theory down to modern times (see [32, 33] for reviews).

But while they have been influential, the extra dimensions of Kaluza-Klein theory are far less radical than Minkowski's, in that they share the length-like character of ordinary space. Philosophically, this represents a return to the Pythagorean prejudice that geometry should deal only in quantities that can be measured with a meter-stick. Given the long history of connections between higher dimensions and unification, it is remarkable that more physicists have not taken Minkowski's example to heart and enlarged the domain of geometry *beyond space and time*.

Among the earliest such proposals were those of William Band (1939) and Osamu Hara (1959) relating x^5 to particle *spin* [34, 35], and that of Yuriĭ Borisovich Rumer, who proposed in 1949 a fifth coordinate based on action via $x^5 = S/mc$ [36]. Rumer applied this idea to what he termed "5-optics" and imposed a restriction (called the "requirement of physical admissibility") similar to Kaluza's cylinder condition [37]. Related ideas were explored more recently by Keller [38].

At its most fundamental, physics deals with dimensions of length [L], time [T] and mass [M], so the most natural physical interpretation for a new coordinate is arguably one related to mass via either $x^5 = Gm/c^2$ or $x^5 = h/mc$. Newton's gravitational constant G (or alternatively Planck's constant h) is thereby promoted to the same dimension-transposing role as c in 4D special relativity. Aspects of this idea were discussed as early as 1967 by de Vos and Hilgevoord [39] and 1974 by

Edmonds [40, 41], but it is most widely associated with Space-Time-Matter theory [42, 43, 44] and its precursors beginning in 1983 [45]. In the context of this theory, the identification of x^5 with rest mass follows naturally from several lines of argument including the fact that the 4D relativistic energy-momentum relation $p^\alpha p_\alpha = E^2 - c^2 p^2 = m^2 c^4$ reduces simply to $p^A p_A = 0$ in 5D (where α ranges over 0,1,2,3 and A ranges over 0,1,2,3,5); and that the 4D free-particle action principle $\delta(\int m\, ds) = 0$ is contained in the simpler 5D one $\delta(\int dS) = 0$ (where $ds^2 = g_{\alpha\beta} dx^\alpha dx^\beta$ and $dS^2 = g_{AB} dx^A dx^B$).

In this interpretation, the fifth component of velocity manifests itself as variable rest mass. The 5D generalization of Eq. (9.2) reads

$$\gamma_5 = \frac{1}{\sqrt{1-(1/c^2)(|d\vec{x}|/dt)^2 \pm (G^2/c^6)(dm/dt)^2}} , \qquad (9.3)$$

where we maintain an open mind with regard to metric signature. The magnitude of the dimension-transposing constant G^2/c^6 explains why no such fifth dimension has yet been detected. The factor $1/c^2$ is already so small that it took until 1908 for time to be recognized as part of space. The factor G^2/c^6 is so much smaller still (by 54 orders of magnitude in SI units) that it is no surprise that a possible variation in, say, the rest masses of elementary particles has yet to be observed. Such variation, if it exists, likely takes place over cosmological scales [46].

Additional timelike dimensions in the context of Kaluza-Klein theory lead to the wrong sign for the Maxwell action, and to the appearance of tachyons (negative-mass eigenstates) in the theory. More generally, additional temporal dimensions raise the spectre of causality violation via closed timelike curves. For these reasons, they have rarely been entertained in the literature. The earliest such proposal may be that of Andrei Sakharov in 1984 [47]. Sakharov considered even numbers of additional compact time dimensions and argued that causality could be preserved for macroscopic processes if the radius of compactification were sufficiently small. This work was further developed by Aref'eva and Volovich [48]. Other "two-time" theories have been propounded by Burakovsky and Horwitz [49], Bars and Kounnas [50], Wesson [51], Kociński and Wierzbicki [52] and Erdem and Ün [53].

Additional kinds of "post-Minkowskian" coordinates have been considered as well [30]. It may be too early to say which, if any, of these ideas will prove to be as useful as Minkowski's in unifying the laws of physics. If and when they do, we may find ourselves echoing his words and announcing that "Spacetime by itself, *and mass by itself*, are doomed to fade away into mere shadows, and only a kind of union of the *three* will preserve an independent reality."

9.3. Interpretation and Experiment

Where is the fifth dimension? The same question would undoubtedly have been asked about the fourth dimension, had Minkowski proposed it *before* 1908. To this day, 4D Minkowski spacetime is sometimes viewed as little more than a geometrical reformulation of SR. But this is misleading. The evidence for time as a fourth coordinate was there long before Einstein, in the form of the *observed Lorentz invariance of electrodynamics*. (In fact, it was this invariance, and not the lack of experimental evidence for an electromagnetic ether, that led Einstein to the discovery of SR itself, as has now been conclusively established by historians of science [54].) That is why Minkowski prefaced his words with the observation that, "these views ... have sprung from the soil of experimental physics, and therein lies their strength."

Is there any experimental evidence, analogous to the Lorentz invariance of electrodynamics, which points to a larger metric structure like that underlying Space-Time-Matter theory? Such evidence would presumably take the form of violations of 4D Lorentz symmetry in the flat-space limit. Consider as a test case the proposition that $x^5 \propto m$, in which case these violations will involve small changes in rest mass, $\dot{m} \equiv dm/dt$ [55]. Existing experiments tell us that departures from 4D SR must be small [56], so Eq. (9.3) can be Taylor-expanded:

$$\gamma_5(v) = 1 + \frac{v^2}{2c^2}\left[1 \mp \left(\frac{G\dot{m}}{c^2 v}\right)^2\right] , \qquad (9.4)$$

where $v \equiv |d\vec{x}/dt|$. The term in square brackets violates Lorentz symmetry. Violations of this kind appear in a kinematical generalization of SR known as Robertson-Mansouri-Sexl (RMS) theory [57]. In RMS theory, the existence of a preferred frame modifies the standard Lorentz transforms such that

$$t = a(v)T + e(v)x \ , \quad x = b(v)(X - vT) \ ,$$
$$y = d(v)Y \ , \quad z = d(v)Z \ , \tag{9.5}$$

where T, X, Y, Z are coordinates in the preferred frame, and the functions $a(v)$, $b(v)$, $d(v)$ and $e(v)$ describe time dilation, length contraction, transverse length contraction and clock synchronization respectively. The function $b(v)$ is the RMS generalization of the Lorentz factor $\gamma(v)$ of SR. Mansouri and Sexl showed on general grounds that

$$a(v) \sim 1 + \alpha \frac{v^2}{c^2} \ , \quad b(v) \sim 1 + \beta \frac{v^2}{c^2} \ , \tag{9.6}$$

where α, β are constants whose values go over to $-\frac{1}{2}$, $+\frac{1}{2}$ in the SR limit and 0,0 in the limit of Galilean relativity. Comparing Eqs. (9.4) and (9.6), we identify

$$\beta - \tfrac{1}{2} = \pm \frac{1}{2} \left(\frac{G\dot{m}}{c^2 v} \right)^2 . \tag{9.7}$$

Experimental constraints on the RMS parameters α, β come from tests of the relativistic Doppler effect, known as the Ives-Stilwell (I-S) experiment; and from the Kennedy-Thorndike (K-T) experiment, a modified form of the original Michelson-Morley experiment with arms of different length. Recent limits are [58, 59, 60]:

$$\left| \alpha + \tfrac{1}{2} \right| \leq 8.4 \times 10^{-8} \quad \text{(I-S)} \ ,$$
$$\alpha - \beta + 1 = 0.0^{+3.7}_{-4.8} \times 10^{-8} \quad \text{(K-T)} \ . \tag{9.8}$$

Combining these expressions, we find that $|\beta - \frac{1}{2}| \leq 1 \times 10^{-7}$. From Eq. (9.7) it then follows that $|\dot{m}| \leq 2 \times 10^{32}$ kg/s (assuming $v \leq c$). The weakness of this constraint is due to the large value of the dimension-transposing constant c^3/G ($= 4 \times 10^{35}$ kg/s in SI units). That is, we "see" this mass-like fifth dimension only at "speeds" $\dot{m}$ close to c^3/G — just as we see Minkowski's fourth dimension only at speeds $\dot{x}$ close to c.

As a purely kinematical model of Lorentz violation, RMS theory has significant limitations. An alternative, fully dynamical theory that contains general relativity and the Standard Model of particle physics along with additional fields that allow for all possible violations of Lorentz symmetry is the Standard-Model Extension or SME [61]. RMS-type models are parametrized within the SME in terms of "rods" and "clocks," observers whose length and time scales can disagree with those of photons when moving at a velocity $\vec{v}$ relative to the preferred frame. In the simplest possible model [62, 63], Lorentz violation is controlled by two new constants, $(c_{\text{clock}})_{\text{TT}}$ and $(c_{\text{rod}})_{\text{TT}}$. To leading order in v,

$$\alpha = -\tfrac{1}{2} - \tfrac{5}{12}(c_{\text{clock}})_{\text{TT}} \;,\quad \beta = \tfrac{1}{2} + \tfrac{7}{12}(c_{\text{rod}})_{\text{TT}} \;. \tag{9.9}$$

Thus in this case, constraints on the RMS parameter $\beta - \tfrac{1}{2}$ in Eq. (9.7) translate directly into constraints on $(c_{\text{rod}})_{\text{TT}}$.

To obtain numerical bounds, we identify our "rods" with particles such as electrons, protons, neutrons or neutrinos, and associate "v" with the speed of the Sun-centered SME frame relative to the average rest frame of photons in the cosmic microwave background, $v \approx 370$ km/s. Experimental limits on $(c)_{\text{TT}}$ are summarized in the SME Data Tables [64]. For electrons they come from synchroton energy losses in accelerators [65], atom interferometry [66], spectroscopy [67], observations of the Crab Nebula flare [68], and clock frequencies [69], implying that $(c_e)_{\text{TT}} > -5 \times 10^{-21}$ and $|(c_e)_{\text{TT}}| < 4 \times 10^{-15}$. For protons and neutrons, they come from atom interferometry [70], nuclear binding energy [71] and atomic fountain clocks [72]. For neutrinos, they come from oscillation experiments like MiniBooNE [73] and Super-Kamiokande [74].

The strongest such limit currently comes from the world's first dual atomic fountain clock and implies that $|(c_p)_{TT}| < 9 \times 10^{-16}$ for protons [75]. Taking this last number as an example, we find from Eqs. (9.7) and (9.9) that:

$$|\dot{m}| = \frac{c^2 v}{G}\sqrt{\frac{7}{6}}|(c)_{\text{TT}}| \lesssim 2 \times 10^{25} \text{ kg/s} \quad . \tag{9.10}$$

Such a result might be consistent with slow variation in particle rest mass on cosmological timescales. Indeed, in this picture mass could *originate* as displacement along x^5, in what might be seen as a geometrical analog of the Higgs mechanism. Related ideas have been explored by Bekenstein [76], Liu and Wesson [77] and Wetterich [78].

In this interpretation, we might expect on dimensional grounds that $|\dot{m}|/M \sim H_0$ where $M \approx \rho_{\text{crit}} L_p^3$ is the mass inside the particle horizon $L_p \approx ct_0$, $\rho_{\text{crit}} \equiv 3H_0^2/8\pi G$ is the critical density, H_0 is the Hubble expansion rate, and $H_0 t_0 = 1$ from observation. We can test this expectation using Eq. (9.10). Combining the foregoing expressions, we find (for protons) that

$$\frac{|\dot{m}|}{M} \approx \frac{2H_0}{3}\left(\frac{v}{c}\right)\sqrt{\frac{7}{6}}|(c)_{\text{TT}}| \lesssim 3 \times 10^{-11} H_0 \tag{9.11}$$

(the constraint for electrons is four times weaker). If the rest masses of elementary particles are indeed varying on cosmological scales, Eq. (9.11) suggests that they are doing so many orders of magnitude more slowly than the cosmic expansion rate. This is a strong constraint, testifying to remarkable progress in the sensitivity of fundamental physics experiments over the past few decades. Of course, many factors need to be considered in interpreting such a result, including the alternative parametrization $x^5 = h/mc$ (Chapter 8). The important point is that Lorentz symmetry must be violated at some level within a 5D-covariant theory, and that another way to test Space-Time-Matter theory may be to situate it systematically within an established framework of experimental constraints like that offered by the Standard Model Extension.

9.4. Conclusion

Space-Time-Matter theory is a minimal extension of general relativity that advances Einstein's dream of unifying the fields of physics with their sources. The fifth dimension in this theory is most naturally (though not necessarily) interpreted as a geometrical parametrization of mass, analogous to Minkowski's fourth coordinate of time. The theory is generally covariant in five dimensions, which means that it must violate 4D conservation laws and be testable in principle. In practice, there is necessarily a degree of ambiguity in constraining the theory, because different metrics in 5D can both describe the same physical situation on 4D spacetime hypersurfaces. The canonical form of the metric is preferred in some ways, and predicts new effects that can be tested through experiments with spinning test bodies, searches for violations of the Equivalence Principle, and observational limits on a time-dependent cosmological "constant." Alternatively, it may be possible to constrain the theory through violations of Lorentz symmetry in the flat-space limit, as explored in a tentative way above.

Space-Time-Matter theory should not be confused with the scale-invariant unified theory of gravity and electromagnetism that was proposed by Hermann Weyl in his 1919 book of the same name, *Raum-Zeit-Materie* [85]. Despite its widely admired formal elegance, that theory came into rapid conflict with experiment and was reluctantly abandoned by its author. Weyl reportedly had his scale-invariant theory in mind when he later confided to Freeman Dyson [86]:

> "My work always tried to unite the true with the beautiful, but when I had to choose one or the other, I usually chose the beautiful."

The empirical status of higher-dimensional Space-Time-Matter theory remains an open question. But like its namesake, it is based on an idea of compelling simplicity and unifying power. The true importance of Weyl's theory was eventually revealed decades after its creation, when it became the model for gauge invariance in modern physics. Time will tell whether Space-Time-Matter theory will someday play a comparable part in the future development of physics.

References

[1]   Clifford, W. K. (1876). On the Space-Theory of Matter, *Proc. Cambridge Phil. Soc.* 2, pp. 157-158.

[2] Misner, C.W., Thorne, K.S. and Wheeler, J.A. (1973). *Gravitation* (W.H. Freeman, San Francisco).

[3] Hartle, J. B. (2003). *Gravity: An Introduction to Einstein's General Relativity* (Addison-Wesley, San Francisco).

[4] Einstein, A. (1922). Geometry and Experience, in *Sidelights on Relativity* (Methuen, London), p. 32.

[5] Einstein, A. (1936). Physics and Reality, *Journal of the Franklin Institute*, Vol. 221, pp. 335 and 370.

[6] Reichenbach, H. (1957). *The Philosophy of Space and Time* (Dover, New York), p. 256.

[7] Bergmann, P.G. (1981). The Quest for Unity: General Relativity and Unitary Field Theories, in Y. Ne'eman (ed.), *To Fulfill a Vision* (Addison-Wesley, Reading, Massachussetts), p. 16.

[8] Lehmkuhl, D. (2014). Why Einstein did not believe that general relativity geometrizes gravity, *Studies in History and Philosophy of Modern Physics* 46, pp. 316-326.

[9] Einstein, A., letter to John Moffat (May 29, 1953), kindly shared by permission of John Moffat and Leopold Halpern. The original German reads: "Bei dieser Sachlage sehe ich mich dazu gedrängt, die logische Einfachheit alleine als Wegweiser zu betrachten unter Verwendung der allgemeinen Relativität. Dies führt mich zu dem Versuch (aber nicht zu der Überzeugung) die Zukunft in einer Feldtheorie (im alten Sinne) zu suchen (Verallgemeinerung der Theorie des Gravitationsfeldes). Wesentlich dabei erscheint mir der Gesichtspunkt, dass die Lagrange-Funktion nicht aus logisch unabhänigen Summanden zusammengesetzt werden darf."

[10] Einstein, A. (1956). *The Meaning of Relativity* (Princeton University Press), p. 129.

[11] Einstein, A. and Grommer, J. (1923). Beweis der Nichtexistenz eines überall regulären zentrisch symmetrischen Feldes nach der Feld-Theorie von Th. Kaluza, *Scripta Universitatis atque Bibliothecae Hierosolymitanarum: Mathematica et Physica* 1, pp. 1-5.

[12] Einstein, A. and Bergmann, P. (1938). On a generalization of Kaluza's theory of electricity, *Annals of Mathematics* 39, pp. 683-701.

[13] Van Dongen, J. (2002). Einstein and the Kaluza-Klein particle, *Studies in History and Philosophy of Modern Physics* 33, pp. 185-210.

[14] Witten, E. (2014). A note on Einstein, Bergmann and the fifth dimension, lecture at the Einstein Centennial Celebration at the Library of Alexandria (June 2005); preprint arXiv:1401.8048 [physics.hist-ph].

[15] Pais, A. (1982). *Subtle is the Lord... The Science and Life of Albert Einstein* (Oxford University Press), p. 152.

[16] Wheeler, J.A. (1957). On the nature of quantum geometrodynamics, *Annals of Physics* (NY) 2, pp. 604-614.

[17] Finkelstein, D. and Rubinstein, J. (1968). Connection between spin, statistics, and kinks, *J. Math. Phys.* 9, pp. 1762-1775.

[18] Halpern, L. (1994). Geometrical structure of gravitation and matter fields, *Int. J. Theor. Phys.* 33, pp. 401-424.

[19] Dzhunushaliev, V.D. (1998). Spherically symmetric solution for torsion and the Dirac equation in 5D spacetime, *Int. J. Mod. Phys.* D7, pp. 909-915.

[20] Trayling, G. and Baylis, W.E. (2001). A geometric basis for the standard-model gauge group, *J. Phys. A: Math. Gen.* 34, pp. 3309-3324.

[21] Vasilić, M. (2001). Consistency analysis of Kaluza-Klein geometric sigma models, *Gen. Rel. Grav.* 33, pp. 1783-1798.

[22] Kalligas, D., Wesson, P.S. and Everitt, C.W.F. (1995). The classical tests in Kaluza-Klein gravity, *Astrophys. J.* 439, pp. 548-557.

[23] Liu, H. and Overduin, J.M. (2000). Solar system tests of higher-dimensional gravity, *Astrophys. J.* 538, pp. 386-394.

[24] Overduin, J.M. (2000). Solar-system tests of the equivalence principle and constraints on higher-dimensional gravity, *Phys. Rev.* D62, p. 102001.

[25] Overduin, J.M., Mitcham, J. and Warecki, Z. (2014). Expanded solar-system limits on violations of the equivalence principle, *Class. Quant. Grav.* 31, p. 015001.

[26] Overduin, J.M., Wesson, P.S. and Mashhoon, B. (2007). Decaying dark energy in higher-dimensional gravity, *Astron. Astrophys.* 473, pp. 727-731.

[27] Overduin, J.M., Everett, R.D. and Wesson, P.S. (2013). Constraints on Kaluza-Klein gravity from Gravity Probe B, *Gen. Rel. Grav.* 45, pp. 1723-1731.

[28] Minkowski, H. (1908), Space and Time in *Hendrik A. Lorentz, Albert Einstein, Hermann Minkowski, and Hermann Weyl, The Principle of Relativity: A Collection of Original Memoirs on the Special and General Theory of Relativity* (Dover, New York, 1952) pp. 75-91.

[29] Goenner, H.F.M. (2008). On the history of geometrization of space-time: from Minkowski to Finsler geometry, lecture at the 414th Heraeus-Seminar: Space and Time 100 Years after Minkowski (September 2008); preprint arXiv:0811.4529 [gr-qc].

[30] Overduin, J.M. (2010). The Experimental Verdict on Spacetime from Gravity Probe B, in V. Petkov (ed.), *Space, Time and Spacetime: Physical and Philosophical Implications of Minkowski's Unification of Space and Time* (Springer, Berlin), pp. 25-59.

[31] S. (1885). Four-dimensional space, *Nature* 31, p. 481.

[32] Halpern, P. (2004). *The Great Beyond: Higher Dimensions, Parallel Universes and the Extraordinary Search for a Theory of Everything* (Wiley, Hoboken, NJ), p. 57.

[33] Overduin, J.M. and Wesson, P.S. (1997). Kaluza-Klein gravity, *Phys. Rep.* 283, pp. 303-378.

[34] Band, W. (1939). Klein's fifth dimension as spin angle, *Phys. Rev.* 56, p. 204.

[35] Hara, O. (1959). A study of charge independence in terms of Kaluza's five dimensional theory, *Prog. Theor. Phys.* 21, pp. 919-937.

[36] Rumer, Y.B. (1949). Action as a space coordinate. I., *Zh. Eksp. Teor. Fiz.* 19, pp. 86-94 (in Russian).

[37] Rumer, Y.B. (1956). *Studies in 5-Optics* (West Siberian Branch of the Academy of Science, Moscow) (in Russian; translated by D.H Delphenich, http://www.neo-classical-physics.info/uploads/3/4/3/6/34363841/rumer_-_studies_in_5-optics.pdf).

[38] Keller, J. 2002. Space-Time-Action Relativity, in *Theory of the Electron. Fundamental Theories of Physics*, Vol. 115 (Springer, Dordrecht: Springer), Ch. 2.

[39] De Vos, J.A. and Hilgevoord, J. (1967). Five-dimensional aspect of free particle motion, *Nucl. Phys.* B1, pp. 494-510.

[40] Edmonds, J.D. (1974). Five-dimensional space-time: mass and the fundamental length, *Int. J. Theor. Phys.* 11, pp. 309-315.

[41] Edmonds, J.D. (1975). Extended relativity: mass and the fifth dimension, *Found. Phys.* 5, pp. 239-249.

[42] Wesson, P.S. *et al.* (1996). A theory of space, time and matter, *Int. J. Mod. Phys.* A11, pp. 3247-3255.

[43] Wesson, P.S. (1999). *Space-Time-Matter* (World Scientific, Singapore).

[44] Wesson, P.S. (2006). *Five-Dimensional Physics* (World Scientific, Singapore).

[45] Wesson, P.S. (1983). A new approach to scale-invariant gravity, *Astron. Astrophys.* 119, pp. 145-152.

[46] Wesson, P.S. (2011). Particle masses and the cosmological 'constant' in five dimensions, preprint arXiv:1111.4698 (2011).

[47] Sakharov, A.D. (1984). Cosmological transitions with changes in the signature of the metric, *Sov. Phys. JETP* 60, pp. 214-218.

[48] Aref'eva, I.Y. and Volovich, I.V. (1985). Kaluza-Klein theories and the signature of space-time, *Phys. Lett.* 164B, pp. 287-292.

[49] Burakovsky, L. and Horwitz, L.P. (1995). 5D generalized inflationary cosmology, *Gen. Rel. Grav.* 27, pp. 1043-1070.

[50] Bars, I. and Kounnas, C. (1997). Theories with two times, *Phys. Lett.* B402, pp. 25-32.

[51] Wesson, P.S. (2002). Five-dimensional relativity and two times, *Phys. Lett.* B538, pp. 159-163.

[52] Kociński, J. and Wierzbicki, M. (2004). The Schwarzschild solution in a Kaluza-Klein theory with two times, *Rel. Grav. Cosm.* 1, pp. 1-19.

[53] Erdem, R. and Ün, C.S. (2006). Reconsidering extra time-like dimensions, *Eur. Phys. J.* C47, pp. 845-850

[54] Norton, J.R. (2014). Einstein's special theory of relativity and the problems in the electro-dynamics of moving bodies that led him to it, in M. Janssen and C. Lehner (eds.), *The Cambridge Companion to Einstein*, Vol. 1 (Cambridge University Press), pp. 72-102.

[55] Overduin, J.M. and Ali, H. (2017). Extra dimensions and violations of Lorentz symmetry, in A. Kostelecký (ed.), *Seventh Meeting on CPT and Lorentz Symmetry* (World Scientific, Singapore), pp. 253-255.

[56] Mattingly, D. (2005). Modern tests of Lorentz invariance, *Liv. Rev. Relativ.* 8, p. 5; https://link.springer.com/article/10.12942/lrr-2005-5.

[57] Robertson, H.P. (1949). Postulate versus observation in the special theory of relativity, *Rev. Mod. Phys.* 21, pp. 378-382.

[58] Mansouri, R. and Sexl, R.U. (1977). A test theory of special relativity: I. Simultaneity and clock synchronization, *Gen. Relativ. Grav.* 8, pp. 497-513.

[59] Reinhardt, S. *et al.* (2010). Test of relativistic time dilation with fast optical atomic clocks at different velocities, *Nature Phys.* 3, pp. 861-864.

[60] Hermann, S. *et al.* (2009). Rotating optical cavity experiment testing Lorentz invariance at the 10^{-17} level, *Phys. Rev.* D80, p. 105011.

[61] Tobar, M.E. *et al.* (2007). Test local Lorentz and position invariance and variation of fundamental constants by searching the derivative of the comparison frequency between a cryogenic sapphire oscillator and hydrogen maser, *Phys. Rev.* D81, p. 022003.

[62] Colladay, D. and Kostelecký, V.A. (1998). Lorentz-violating extension of the standard model, *Phys. Rev.* D58, p. 116002.

[63] Kostelecký, V.A. and Mewes, M. (2002). Signals for Lorentz violation in electrodynamics, *Phys. Rev.* D66, p. 056005.

[64] Kostelecký, V.A. and Mewes, M. (2009). Electrodynamics with Lorentz-violating operators of arbitrary dimension, *Phys. Rev.* D80, p. 015020.

[65] Kostelecký, V.A. and Russell, N. (2011). Data tables for Lorentz and CPT violation, *Rev. Mod. Phys.* 83, p. 11.

[66] Altschul, B. (2010). Laboratory bounds on electron Lorentz violation, *Phys. Rev.* D82, p. 016002.

[67] Hohensee, M.A. *et al.* (2011). Equivalence principle and gravitational redshift, *Phys. Rev. Lett.* 106, p. 151102.

[68] Hohensee, M.A. *et al.* (2011). Limits on violations of Lorentz symmetry and the Einstein equivalence principle using radio-frequency spectroscopy of atomic dysprosium, *Phys. Rev. Lett.* 111, p. 050401.

[69] Hohensee, M.A., Müller, H. and Wiringa, R.B. (2013). Equivalence principle and bound kinetic energy, *Phys. Rev. Lett.* 111, p. 151102.

[70] Stecker, F.W. (2014). Limiting superluminal electron and neutrino velocities using the 2010 Crab Nebula flare and the IceCube PeV neutrino events, *Astropart. Phys.* 56, pp. 16-18.

[71] Dzuba, V.A. and Flambaum, V.V. (2017). Limits on gravitational Einstein equivalence principle violation from monitoring atomic clock frequencies during a year, *Phys. Rev.* D95, p. 015019.

[72] Wolf, P. *et al.* (2006). Cold atomic clock test of Lorentz invariance in the matter sector, *Phys. Rev. Lett.* 96, p. 060801.

[73] Katori, T. (2012). Tests of Lorentz and CPT violation with MiniBooNE neutrino oscillation excesses, *Mod. Phys. Lett.* A27, p. 1230024.

[74] Abe, K. *et al.* (2015). Test of Lorentz invariance with atmospheric neutrinos, *Phys. Rev.* D91, p. 052003.

[75] Pihan-Le Bars, H. *et al.* (2017). Lorentz-symmetry test at Planck-scale suppression with nucleons in a spin-polarized ^{133}Cs cold atom clock, *Phys. Rev.* D97, p. 075026.

[76] Bekenstein, J.D. (1977). Are particle rest masses variable? Theory and constraints from solar system experiments, *Phys. Rev.* D15, pp. 1458-1468.

[77] Liu, H. and Wesson, P.S. (2000). On the Klein-Gordon equation in higher dimensions: are particle masses variable? *Gen. Rel. Gravit.* 32, pp. 583-592.

[78] Wetterich, C. (2014). Variable gravity universe, *Phys. Rev.* D89, p. 024005.

[79] Weyl, H. (1952). *Space Time Matter* (Dover).

[80] Chandraskehar, S. (1990). *Truth and Beauty: aesthetics and motivations in science* (University of Chicago Press), p. 65.

Appendix

Paul Wesson (1949–2015)

A.1. Nottingham

Paul Stephen Wesson was born in Nottingham, England on September 11, 1949. His father was an automobile mechanic and his mother a homemaker. Paul was later to recall that there were no books in the house. Favourite pastimes included boating on the river and collecting newts. From age 11 to 16, he and his younger brother Clive attended the nearby Chandos Street Boys' School (Fig. A.1), where discipline was prized over ambition. One alumnus recalled: "We used to say they should have a sign above the door: 'Abandon hope all ye who enter here.' Best you could hope to be was a plumber, electrician, or mechanic." This working-class background formed and deeply stamped Paul's character, a fact remarked on by nearly all who knew and worked with him throughout his life. Here he developed his capacity for hard work, his strong sense of personal loyalty, his lack of pretension (and suspicion of pretense in others), his cheerful willingness to take on established authority, and his wickedly dry sense of humor.

Paul shared an especially close childhood bond with his maternal grandmother, a Nottingham shop owner who was seen as the professional member of the family. Perhaps under her influence, he excelled at school, especially in English and physics. He climbed from 6th to 1st place in his class within the first year, and remained in top spot thereafter. He participated eagerly in a school-sponsored pen-pal program with a partner school in Nanaimo,

Fig. A.1: School days (1962)

217

Fig. A.2 (left): Paul (front row center) with fellow participants in a naval training program in Portsmouth (1965). Fig. A.3 (right): Paul in Nottingham with his first motorcycle, a 250cc Matchless Pathfinder (1966). He later traded up for a 500 cc BSA, and eventually a 350 cc Triumph Tiger, which he remembered as "the best of the lot."

British Columbia. Through this long-distance friendship Paul determined to "escape the claustrophobia of England" (as he later put it) and emigrate to Canada someday. As luck would have it, he did eventually settle on an island near Nanaimo during the final decade of his life.

Paul had no clear idea of a career when he finished high school in 1966. A keen member of the school sailing club, he briefly considered joining the Royal Navy (Fig. A.2). (According to a story he liked to tell later on, he was accepted into the officer training program in Portsmouth but lasted only seven days.) He took a summer job landscaping and used the proceeds to buy the first in a series of used motorcycles (Fig. A.3). But within a year he had decided on science. From 1967-68 he obtained the necessary qualifications at a Nottingham vocational school, the Arnold and Carlton College. He excelled in geology and was accepted into a BSc program run by the University of London at what was then Portsmouth Polytechnic (now the University of Portsmouth).

A.2. Portsmouth and London

Paul was initially torn between geology and physics. Alan Warrington, Paul's friend and roommate during his Portsmouth years, recalls that he was strongly influenced by the head of the Geology Department, Brian Walton, who founded a local branch of the Humanist society which both students joined (Fig. A.4). He earned money roaming the countryside on his motorcycle, collecting rock samples for a geology laboratory supply company. And he spent his first summer as an undergraduate in the Portsmouth public library, where he wrote a review article questioning the evidence for continental drift.

Fig. A.4: Paul (right) with friend and roommate Alan Warrington outside the physics department of Portsmouth Polytechnic (1970)

This was a bold if doubtful position to take in 1969; the reality of plate tectonics had been accepted by most of the world's experts by the time of the climactic 1967 annual meeting of the American Geophysical Union, if not before. Nevertheless, Paul's paper was accepted for publication in the respected *Quarterly Journal of the Royal Astronomical Society* the following year [1]. It was apparently refereed by the distinguished Cambridge mathematician and theoretical astronomer Raymond Lyttleton, a drift skeptic. In the summer of 1970, Paul traveled to Cambridge to meet with another lifelong opponent of plate tectonics, the eminent mathematician and geophysicist Sir Harold Jeffreys. Encouragement from prominent scientists such as these must have been a heady experience for the 20 year-old student, who published a second paper on the subject in *Nature* while still an undergraduate [2].

Three factors, however, conspired to divert Paul's career away from geology. First was the head of the Portsmouth Physics Department, Jim Skane, a charismatic figure who took a personal interest in Paul and convinced him to switch out of Applied Physics and into a smaller and more theoretical "Special Physics" program overseen by the University of London. Few of the students who enrolled in this program completed it. But Paul thrived, drawing special notice in Electricity and Magnetism, where he earned a rare "A++" grade. He graduated from the University of London with First Class Honours in 1971, and was one of a dozen students to be invited by Britain's Astronomer Royal, Sir Richard Woolley, to

Fig. A.5: Participants in the 1971 astronomy summer school at Herstmonceux Castle, including (from left to right) Stella Harris, Paul Wesson, Jonathon Holmes, Susan Jones, Paul Kienewicz, Ian Philpott, Mike Hawkins (Image courtesy Bernard Carr).

spend that summer conducting research at the Royal Greenwich Observatory, then located at historic Herstmonceux Castle.

The summer at Herstmonceux was the second factor, awakening Paul's interest in astronomy despite a damp east Sussex climate that was hardly conducive to observation (he remarked later that "the domes could be seen from afar, floating like bubbles above the mists of the Pevensey marshes"). He studied stellar spectra, published a third article on plate tectonics [3] and formed lifelong friendships with the other students (Fig. A.5). One of those students was Bernard Carr, who later recalled: "It was an idyllic summer and my first memories of Paul are also idyllic. Because of our wide range of rather unconventional interests, we were kindred spirits, although I was more conservative than him and had shorter hair. One of our first experiences was having dinner together in a pub. Apparently I stepped in to save a bug which the waitress was trying to kill and this small act of kindness convinced him I was a good person. I was a Buddhist at the time, and it's ironic that a bug cemented our friendship."

In the fall of 1971, Paul was accepted into Cambridge University to read Part III of the Mathematical Tripos. His experiences here would be the third and final factor behind his change of field. He continued his friendship with Bernard, who was already at Cambridge. They talked about extra dimensions, attended lectures on cosmology by George Ellis, improvised together on the pipe organ of St. John's College, and passed the grueling exam that qualified them for the doctoral program in some area of mathematical physics. Bernard went on to do a PhD with Stephen Hawking, while Paul took a break

Fig. A.6: Paul in Afghanistan (1972)

from studies, which would also turn out to mark his final break with geophysics. Together with five colleagues, he traveled overland to northeastern Afghanistan as part of the 1972 Cambridge Hindu Kush Expedition, whose purpose was to explore unclimbed peaks and report back on seismic activity in the region (Fig. A.6). This experience had a profound impact on the course of Paul's life [4]. Geology had been his first love, but he now saw it as scientifically too descriptive and mathematically too simple. He would go on to write several more short papers on geological topics [5, 9, 14, 20, 44, 83], but his interest from this time on shifted decisively toward astrophysics and cosmology.

A.3. Cambridge

Back at Cambridge at the end of 1972, Paul found the departments of astronomy and theoretical physics in turmoil (Fig. A.7). Conflict had boiled over between Sir Fred Hoyle, Plumian Professor and Head of the Institute of Theoretical Astronomy, and the radio astronomer Sir Martin Ryle, who had succeeded Woolley as Astronomer Royal. Convinced that he was being forced out, Hoyle resigned, to be followed by professors Jayant Narlikar and Chandra Wikramasinghe. Paul had approached all four men as possible research supervisors, and his hopes were now dashed.

He and a half dozen other students found themselves without
mentors until the arrival in
1973 of a new Plumian Professor, Martin Rees.

Rees (later Sir Martin,
and eventually Lord Rees) took
on all the orphaned students
himself, but was often stretched
for time. During his first three
years as a doctoral student, Paul
published more than twenty
articles on a wide range of
subjects, including the electromagnetic properties of astrophysical dust [6, 16, 21], geophysical implications of cos-

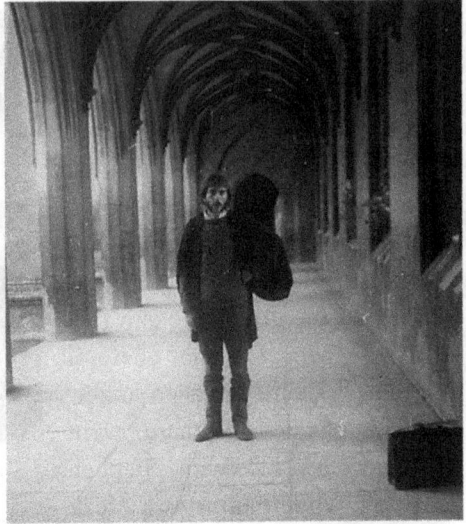

Fig. A.7: Paul in New Court, St. John's College, Cambridge University (1972)

mology with a variable gravitational constant [7], galactic dynamics
[8, 10], hierarchical cosmology [11, 12, 17, 18, 19, 22, 26, 28], galaxy
clustering [13, 15, 25] and astronomical statistics [23, 24]. All were
single-author. He was elected to membership in Britain's Royal
Astronomical Society in 1974.

Paul had not forgotten his youthful fascination with Canada and
the wide, northern spaces of the world, and he obtained permission from
Cambridge to travel abroad for a series of extended visits. The first, in
1976-77, was to Queen's University in Canada (Fig. A.8), where he
collaborated with Norwegian astrophysicist Aksel Lerman on a diverse
array of topics including dust [27, 33], magnetohydrodynamics in jets
[29], dark matter in galaxy clusters [30, 32, 34, 36], the formation of small
solar-system bodies [38] and self-similarity in rotating solutions of
Einstein's equations [39]. This last paper foreshadowed Paul's gradual
shift toward gravitational theory.

Back at Cambridge, he published his first book [37], an expanded
version of his thesis for Part III of the Mathematical Tripos. He continued
to work on galactic dark-matter halos [43] and the role of dust in planetary
formation [48], but focused increasingly on relativistic cosmology [40-

42, 45, 49-52]. In 1978 he embarked on a systematic study of self-similarity in general relativity with Canadian astrophysicist Richard Henriksen, also at Queen's University [46, 47]. Richard later remembered Paul as "a good, talented man who was never satisfied with the ordinary ... His energy and enthusiasm for life were very apparent ... Paul was a prodigious worker. He could write a manuscript in a day and a night once the substance was agreed upon. His ideas were always an honest attempt to solve big questions. I [also] remember his fondness for a beer in the faculty club ... I

Fig. A.8: Paul in British Columbia, fishing with the family of his Canadian childhood penpal (1976)

would find him sometimes there to watch *Batman* on TV. He had this mixture of talent and childish enthusiasm."

A.4. Norway and Canada

By 1979, Paul was thirty years old and had published over fifty papers. He obtained his Ph.D. from Cambridge that year with a doctoral thesis based on his cumulative work on galactic astrophysics to that point. He then traveled abroad for two more extended research trips. The first, supported by the Royal Astronomical Society, took him to Canada's west coast, where he worked on dark matter [55] and variable gravity [56] at the University of British Columbia in Vancouver. The second was a NATO postdoctoral fellowship, which Paul chose to bring to the University of Oslo in Norway. Here he published his second book, *Gravity, Particles and Astrophysics* [53], and worked on the usual range of questions including the formation of the solar system [54, 58], dust [60] and cosmic rotation [62]. He also began two projects that were to bear significant fruit later on: one on dimensional analysis in cosmology [59] and another on extragalactic background light [57]. The latter would eventually develop into a major collaboration with Norwegian cosmologist Rolf Stabell.

In his spare time, Paul took language classes, becoming proficient enough that Norwegian became his preferred language for the many scribbles he entered into the margins of his books and articles. He met and married Ellen Stauborg. Bob Stallman, a fellow language student, cross-country skiing partner and close friend from this time, later recalled: "What I appreciated most about him was his loyalty. In spite of the geographical distance between us, he always kept in touch. ... I also found Paul a very down-to-earth person with considerable humility. A man with two degrees and 300 scholarly articles, [he seemed] more like a truck driver than a professor."

A.5. Edmonton

In the fall of 1980, Paul accepted a position as Assistant Professor at the University of Alberta in Canada. His years in Edmonton were quiet but critical ones, in which he laid the groundwork for the main new ideas which would go on to occupy him for the remainder of his career. Some of these traced their origin back to Paul's time in Norway, like a paper titled "Clue to the unification of gravitation with particle physics" [61], whose acknowledgments reveal that "The ideas described above were conceived in the SAUNA division of Toyenbadet (Oslo, Norway)." He reconsidered the case for large-scale homogeneity [64] and proposed what he termed an "improved standard cosmology" based on embedding local spherically-symmetric regions in a globally homogeneous background [63, 70]. He explored the relationship between gravity and angular momentum [68, 69, 75], the dynamics of galaxy clustering [71, 74] and the then-new theory of cosmic inflation [73, 76]. But Paul's most consequential work during this period originated as a program to re-express the laws of gravity in scale-invariant form [65-67, 72]. This quest led him in 1983 to a paper titled "An embedding for general relativity with variable rest mass" [78], an early version of what would eventually develop into five-dimensional Space-Time-Matter theory.

While in Alberta, Paul traveled regularly to the Dominion Astrophysical Observatory in neighbouring British Columbia, and he discussed some of the above projects with Canadian colleagues including Sidney van den Bergh [69], Paul Hickson [74], Werner Israel [75] and

Richard Henriksen [76]. But mostly he worked alone. For relaxation he went climbing in the Rocky Mountains and developed a love of ice hockey. He cheered on Wayne Gretzky and the Edmonton Oilers during home games with newly born daughter Amanda on his lap, and played regularly himself as part of an old-timers' league. Nevertheless, Paul would later allude to a feeling of scholarly isolation in Edmonton. This situation was resolved, as he later put it, "by a chance meeting with another astronomer at the observatory in Victoria and a talk over a glass of beer while waiting for the skies to clear." In 1984, Paul accepted a new position as Associate Professor at the University of Waterloo in Ontario (Fig. A.9), where he would be based for the remainder of his career.

Fig. A.9: Paul in the University of Waterloo observatory. Taken from a campus newspaper story which goes on to refer to "the 10 thousand million years the university is known to have been in existence," showing how Paul had enhanced the prestige of his new institution (*University of Waterloo Gazette*, "By light of stars physicist ponders ultimate questions," January 24, 1990).

A.6. Waterloo

Paul's subsequent research can be usefully divided into two periods: before and after his first sabbatical in California from 1990-91. The years 1984-90 were transitional ones in which he continued to develop some of his earlier projects on topics such as the dynamics of galaxy clusters [77, 84, 91], the origin of angular momentum in the Solar System [79] and the implications of dimensional analysis for astronomy [81, 102]. Three themes, however, now began to dominate his scientific output. The first was mathematical cosmology. Paul found new solutions of Einstein's equations that could describe successive stages in the history of the early Universe [80, 90, 99, 100, 104, 109], including one that evolved smoothly from an initially empty Minkowski state into the standard expanding Friedmann-Lemaître-Robertson-

Fig. A.10: Paul (right) discussing the big bang with Pope John Paul II at the Vatican (1985)

Walker model with no big bang [85, 89, 94]. These ideas attracted the interest of Pope John Paul II, who invited Paul to the Vatican for an audience along with about twenty other cosmologists (Fig. A.10). In his contribution to the proceedings of this meeting, Paul lamented what he saw as the overly dogmatic attitude of some of his professional colleagues [92]. By contrast, he defended the more speculative cosmologies of earlier thinkers like Arthur Eddington, Paul Dirac, George Gamow, Arthur Milne, William McCrea and Fred Hoyle in a letter to *Physics Today*, noting that "they may not always be right. But they have a place in physics: it is to ask fundamental questions" [110].

This trip to Rome also brought out a lighter side of Paul that delighted in puncturing pretensions. Alan Coley, a fellow Canadian participant in the Vatican conference, later recalled how the group was taken to visit some historic churches. After hearing someone murmur, "What a beautiful façade!" Paul turned to Alan and said loudly enough for everybody to hear: "Façade — that's the front, ain't it?"

A second emerging theme in Paul's research during this time grew out of his postdoctoral work on the extragalactic background light or EBL [57, 86, 97]. Together with Rolf Stabell and Norwegian graduate student Knut Valle, Paul found an innovative way to quantitatively compare the bolometric intensity of the EBL in cosmological models with and without expansion, allowing him to prove in 1987 that expansion plays only a minor role in determining the brightness of the night sky [93, 96]. This discovery received wide notice in the field, dispelling lingering myths about Olbers' paradox and replacing them with a renewed awareness of the profound link between the brightness of the night sky and the age of the universe [101, 103, 105]. Largely on the basis of his achievements in cosmology, Paul was promoted to Full Professor in 1988.

The third subject that began to dominate Paul's work in Waterloo was his scale-invariant embedding for general relativity, which he now referred to as "Kaluza-Klein cosmology with variable rest mass" [82, 88, 95, 98]. The new theory attracted interest from a diverse group of researchers including Alan Coley, Sujit Chatterjee in India, Takao Fukui in Japan and Chilean-born Jaime Ponce de Leon in Venezuela (the beginning of what would eventually become known as the 5D Space-Time-Matter Consortium). In response to criticism from Norwegian physicists Øyvind Grøn and Harald Soleng, Paul acknowledged that the theory was not, in fact, generally scale-invariant; and emphasized that while it took its mathematical inspiration from Kaluza, it was otherwise physically distinct from traditional Kaluza-Klein theory [108]. In a brief review in 1990, he speculated for the first time "that the 5D theory may be complete without an explicit energy-momentum tensor ... the extra terms present in the equations $G_{AB} = 0$ in 5D may play the role of the matter terms that appear on the right-hand sides of the equations $G_{\mu\nu} = T_{\mu\nu}$ in 4D. If this interpretation is correct, *the field equations of the new theory can be taken to be $G_{AB} = 0$ with generality*" ([106], emphasis added). The ramifications of this idea would occupy Paul for the rest of his life.

A.7. Teacher and Mentor

It is perhaps appropriate here to say something about Paul as a teacher and research mentor during these years. In the classroom, he had a dry, detached lecturing style that was not particularly engaging. The impression one got was that he knew the subject itself was fascinating enough to draw in the right students, and he did not need to dress it up. It was also plain that his mind was preoccupied with research, and he did not ask for much in terms of homework or tests, leaving it to students to approach him if they were interested. Those students who *did* approach him, however, were richly rewarded. He tended to form his judgments quickly, whether rightly or wrongly; and once he had made up his mind that a student had promise, he would go to unusual lengths on their behalf. One such student had already graduated and gone to work in a good-paying industry job, when he received a long-distance telephone call from Paul urging him to come back to school and conduct research as a graduate student. (He did.)

Over the course of his career, Paul supervised a total of 14 Masters students, 13 doctorates, and 17 postdoctoral fellows. He was an ideal research mentor: available to guide when needed, but ready to grant a dangerous degree of autonomy if he sensed that a student was ready. More often than not, he encouraged his students to be the first authors on joint publications. Supportive and generous, he was always there to answer a phone call from a panicked student at a conference halfway around the world, to mail a letter of recommendation on short notice, to share the last of his grant money, or even to offer a student free boarding in time of need. Indeed, he often struck his research students as more like a roommate than a professor. A typical anecdote comes from Andrew Billyard, who shared a 5000-mile return road trip with him from Waterloo to Stanford for a conference: "It was day one of driving through the Nevada desert, midafternoon. I had begun to fall asleep in the passenger seat, to be suddenly jolted awake as the car veered. I looked over to see what was wrong, only to see Paul trying to roll a cigarette. To accomplish this, he was steering the 1970s-era Buick station wagon with his knees as he transferred the contents of the tobacco pouch on his lap to the papers in his hand. I said (somewhat concerned for our lives), "Paul, would you like me to do that for you?" To which he replied with his low and steady tone, "Sure, if you like; I didn't want to disturb your slumber."

A.8. Berkeley and Stanford

In 1990, Paul co-organized an international conference on gravitation in Banff, British Columbia (Fig. A.11), the proceedings of which were published as his third book [111]. Fred Cooperstock, a fellow Canadian relativist who also participated in this meeting, later recalled: "We saw in each other kindred scientific spirits, which was the foundation of our

Fig. A.11: Paul (second from left) with several other participants in the 1990 Banff Institute on Gravitation in British Columbia, Canada (from left to right): John Moffat, Takao Fukui and Fred Cooperstock (photo courtesy Takao Fukui).

relationship. I admired Paul's scientific independence, his drive to pursue his scientific ideas courageously and with determination, wherever they led."

Paul then traveled to California for a yearlong sabbatical, which he divided between the Space Sciences Laboratory at the University of California, Berkeley, and the Hansen Experimental Physics Laboratory at Stanford University. It was here, on the eucalyptus-scented slopes of the East Bay Hills overlooking San Francisco Bay, with eagles occasionally soaring below his office window, that Paul assembled the main ingredients of Space-Time-Matter theory. With two exceptions, this project would occupy him for the rest of his career.

The first exception was astrobiology. It accounted for only a small fraction of Paul's scientific output but was important to him nonetheless. His interest in this subject was undoubtedly inspired by Fred Hoyle, whose influence can also be discerned in Paul's science-fiction novels [209, 220, 274] and short stories [281]. He was also stimulated by discussions with Stuart Bowyer, his host at Berkeley and the moving force behind SERENDIP, one of the longest-running searches for extra-terrestrial intelligence (SETI). Paul's position on SETI was initially pessimistic [107], but softened in later years as he worked with graduate student Jeff Secker and biophysicist James Lepock to investigate the possibility (known as panspermia) that life might have been able to propagate through the vast reaches of interstellar space [142, 157, 169]. He eventually concluded that it could not, but left open the possibility that its *information content* might, a hypothesis he provocatively dubbed "necropanspermia" [275]. Fig. A.12 shows Paul

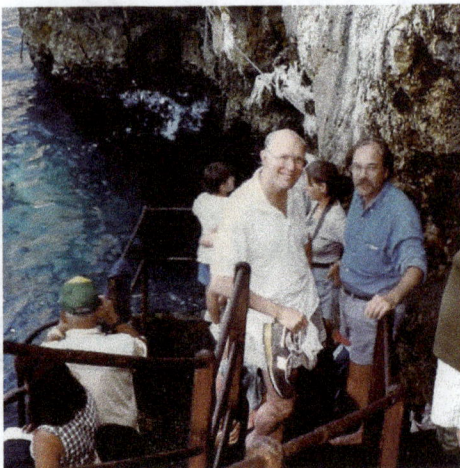

Fig. A.12: Paul (right) with astronomer John Rather at a bioastronomy conference in Capri (1996)

with NASA astronomer John Rather at a bioastronomy conference on the Italian island of Capri in 1996. High waves had prevented their tour boat from visiting the famous Blue Grotto. Undaunted, the two scientists stripped down to their shorts and swam in on their own.

The second exception grew out of Paul's work on bolometric EBL intensity with Rolf Stabell [93]. At Berkeley, he generalized those calculations to spectral EBL intensity [112] and realized that the same equations could also be used to model background radiation from more exotic processes, such as the decay or annihilation of dark matter. Data on the diffuse background at various wavelengths might then constrain or even rule out some dark-matter candidates. Together with Bowyer and Waterloo graduate student James Overduin (who accompanied him to Berkeley), Paul set new limits on decaying vacuum energy [129] and neutrinos [131, 171, 186, 190]. He and James later extended the same method to axions [133], WIMPs [168], primordial black holes and others [123, 127, 175, 194, 233], summing up their results in two books, *Dark Sky, Dark Matter* in 2003 [219] and *The Light/Dark Universe* in 2008 [252]. In 2003, Paul was awarded a D.Sc. (second doctorate) from the University of London "for fundamental contributions to cosmology including background radiation and dark matter".

Paul was involved with vacuum energy in another way at about this time, one with a distinctly Californian twist. Beginning in 1991, he wrote several papers exploring the cosmological significance of quantum zero-point energy [113, 115, 125]. He was then invited to serve from 2000-2002 as Chairman of the Science Advisory Board for the California Institute for Physics and Astrophysics, a privately funded Palo Alto think-tank led by astrophysicist Bernard Haisch, one of whose goals was to study the feasibility of harnessing the energy of the vacuum. Paul assembled a board of eminent colleagues (including Mirjam Cvetic, Andrei Linde, Bahram Mashhoon and Wolfgang Rindler) and wrote a white paper detailing open questions and recommendations for future research [198]. Unfortunately, the institute's financial basis melted away amid the collapse of the Silicon Valley dot-com bubble, leaving a rueful Paul to reflect that "privately-funded institutions for research *can* succeed, but they need very solid financial arrangements, and (more importantly perhaps) they need broad scientific foundations" [255].

Astrobiology and dark matter/energy aside, essentially all Paul's research after 1990 was devoted to Space-Time-Matter theory, also variously referred to at different points as induced-matter theory, non-compactified Kaluza-Klein theory, Kaluza-Klein gravity and five-dimensional relativity. By the end of his life this work would account for nearly two thirds of Paul's more than 300 publications and garner more than 5000 of his approximately 7000 career citations. It is not necessary to review this material in detail, since Paul did so himself over the course of numerous review articles [154, 167, 255, 300] and books including *Space-Time-Matter* in 1999 [184], *Five-Dimensional Physics* in 2006 [241], and the present volume, his last word on the subject. The account that follows is thus kept intentionally brief.

A.9. Space-Time-Matter Theory

The core idea remained unchanged: matter and energy in 4D could be regarded as manifestations of geometry in 5D. A key tenet of the theory continued to be covariance in 5D, not 4D; meaning that physics had to be allowed to depend on the fifth coordinate in principle (thus relaxing the "cylinder condition" of traditional Kaluza-Klein theory). As Space-Time-

Matter theory developed, however, the fifth coordinate was no longer necessarily identified physically with rest mass from the outset, and the moniker "variable-mass theory" was gradually dropped.

A first task in the subsequent evolution of the theory was to revisit the classical tests of general relativity (plus the geodetic precession test) using a 5D generalization of the Schwarzschild metric. This work was carried out at Stanford beginning in 1990 by Paul together with graduate student Dimitri Kalligas and Francis Everitt, Principal Investigator of the Gravity Probe-B experiment [146], with contributions from Paul Lim and graduate student James Overduin at the University of Victoria [122, 124, 151]. Lim, who lived on a sailboat, disappeared during a voyage home from Hawaii to British Columbia in the summer of 2016 and is presumed to have drowned. His probable death at the age of 68 was a tragic loss.

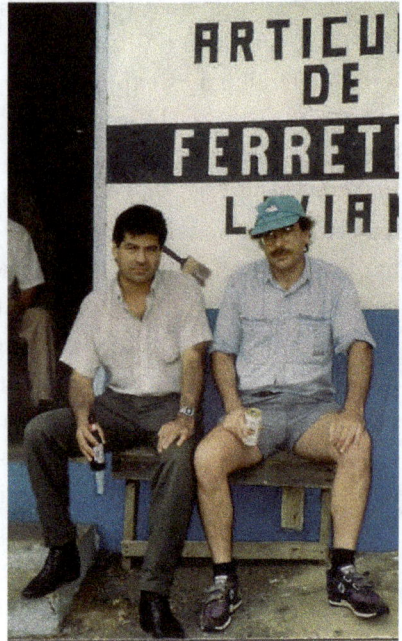

Fig. A.13: Paul (right) and Jaime Ponce de Leon in Puerto Rico (1995)

Another early milestone in the growth of Space-Time-Matter theory was the discovery by Paul and Jaime Ponce de Leon of 5D cosmological solutions that reduce to standard Friedmann-Lemaître-Robertson-Walker models on 4D hypersurfaces [104], and then (in 1992) of the general form of the 4D induced-matter energy-momentum tensor, Eq. (2.26) [128]. This result would provide the impetus for much subsequent progress in the field. The two continued to collaborate closely after Jaime obtained a faculty position in Puerto Rico (Fig. A.13), eventually co-authoring 14 papers.

A third key step was the introduction in 1994 of the "canonical form" of the metric, Eq. (2.6), by Paul together with Bahram Mashhoon and Hongya Liu at the University of Missouri-Columbia [144]. This is a way of writing the 5D metric that uses the available coordinate degrees of freedom to clarify the physics without loss of algebraic generality (similar to the synchronous gauge in ordinary 4D relativity). In the canonical gauge, the physical identi-

Fig. A.14: Paul (right) in discussion with Bahram Mashhoon at a winery overlooking the Missouri river (2003)

fication of the fifth coordinate with mass is recovered. The 5D equations of motion take on a particularly transparent form, reducing to their 4D counterparts when the 4D part of the metric is independent the of fifth coordinate. If the 4D part of the metric does depend on the fifth coordinate, then new effects are predicted that can be tested through gyroscopic experiments [155, 291], searches for violations of the weak Equivalence Principle [191, 225] and observational limits on a time-dependent cosmological "constant" [231, 251].

Paul worked on these subjects with Bahram, Hongya and others over the course of multiple research visits (Fig. A.14), and summed them up in two editions of his book *Space-Time-Matter: Modern Kaluza-Klein Theory* [184]. Bahram, a co-author on 13 papers, praised Paul's "scholarly approach, high standards, rigorous methods of investigation and uncompromising scientific attitude," remarking that "these qualities render any collaborative endeavor with him fruitful as well as a learning experience." Hongya, who subsequently obtained a faculty position at Dalian University in China, became Paul's most prolific co-author, with 33 shared articles between them. Quiet, hardworking, and dedicated above all, his death from cancer in 2008 at the age of 61 was a personal as well as professional blow [260].

A fourth important stage was marked by the theory's increasing application to problems involving particle physics and wave mechanics, as well as astrophysics and cosmology. Paul carried out this work during the 1990s with a succession of talented graduate students at Waterloo, including Andrew Billyard [152], Bill Sajko [189], Sanjeev Seahra [205], Tomas Liko [228], Dan Bruni and others. He summed up the results in a new book, *Five-Dimensional Physics: Classical and Quantum Consequences of Kaluza-Klein Cosmology* [241].

A fifth breakthrough occurred with the realization by Paul and Sanjeev in 2003 that the theory's mathematical foundation had been guaranteed all along by a result in differential geometry known as Campbell's theorem, which proves that Einstein's field equations in 4D ($G_{\mu\nu} = T_{\mu\nu}$) can always be smoothly (if locally) embedded in the 5D Ricci equations ($R_{AB} = 0$) [224, 240]. This connection established the theory more securely and led to new links with other approaches to higher-dimensional unification, including embedding and membrane theory. Reza Tavakol, a mathematician and astronomer who was instrumental in rediscovering Campbell's theorem in the 1990s, knew Paul through a shared connection with the University of London and later recalled: "My memories of him are all of a humorous, kind and caring man and a loyal friend."

A.10. Victoria and Gabriola Island

As he reached his mid-fifties, Paul had over 200 publications to his credit and showed no sign of slowing down. He spent increasing periods of time away from Waterloo, searching for tests of Space-Time-Matter theory in discussions with experimentalists and astronomers at Stanford University's Hansen Experimental Physics Laboratory, the Canadian Institute for Theoretical Astrophysics in Toronto, and the Herzberg Institute of Astrophysics in Victoria (the rechristened Dominion Astrophysical Observatory). These trips kept him connected with a worldwide network of lifelong friends and fellow scientists (Fig. A.15). He joined former students and colleagues in California for the launch of the Gravity Probe-B test of general relativity (Fig. A.16). One of those colleagues, Ron Adler, reminisced: "I will miss Paul for his original ideas on gravity and geometry, and even more for his good humor in chats about how the Universe really works, many of them over a friendly beer. He was one of a kind, as his many friends all know."

Paul co-authored a book on cosmology with physicist and popular science writer Paul Halpern [242]. Halpern was one of many people to dine with Paul at his favourite establishment, Waterloo's Duke of Wellington pub. He remembered Paul as a kind and gracious host but also one who never really relaxed, writing at one point: "I am presently on vacation, which for me means working less than normal. I am one of those people who have to work to feel justified."

Fig. A.15: Paul (right) with Bernard Carr during a stay at the Canadian Institute of Theoretical Astrophysics in Toronto, where both were frequent visitors (2001).

Paul suffered during the last two decades of his life from a combination of chronic pancreatitis and diabetes. He was by this time a divorced single parent of three young children (Amanda, Emily and Jasper). Still affiliated officially with the University of Waterloo, he was able to obtain an early retirement from teaching while continuing his research. Good fortune eventually intervened in the person of a new life partner, Patricia Lapcevic, a hydrogeologist. In 2004, the couple welcomed a son, Sterling, Paul's fourth child (Fig. A.17), and realized Paul's childhood dream by moving together to Canada's west coast. Living beside the sea on Gabriola Island, Paul found a measure of personal contentment at last. He became a lifetime member of the cricket club in nearby Victoria, whose distinctly English flavour he found congenial. He could often be found on the beach, smoking his pipe, collecting driftwood, and walking together with a group of retired men from the local pub, most of whom never knew that he was a professor of physics.

Paul's research continued unabated, much of it at the Herzberg Institute of Astrophysics, where he took up a visiting position beginning in 2007. His scientific and literary output eventually totaled 305 publications (including 267 articles, 25 preprints, 9 books, 3 science-fiction novels, and a collection of science-fiction short stories). His last book, *Weaving the Universe: Is Modern Cosmology Discovered or Invented?* [273], was a summing-up of sorts, in which Paul reviewed the status of higher dimensions, time, matter, religion, and science; reaching the conclusion that all of these are, to a larger extent than commonly supposed, products of the human mind. In a particularly remarkable passage he argued on the

Fig. A.16: Paul (right) with Ron Adler (left) and former student James Overduin (center) near Vandenberg Air Force Base in California, following the successful launch of Gravity Probe-B (2004).

basis of Space-Time-Matter theory that death itself should be seen as a phase change, not an endpoint; and should therefore not be feared: "We obtain a simple model wherein existence is described by a hypersurface in a higher-dimensional world, with two modes of which one is growing and is identified with corporeal life, while one is wave-like and is identified with the soul, the two modes separated by an event which is commonly called death. Whether one believes in a model like this which straddles physics and spirituality is up to the individual. (In this regard, the author is steadfastly neutral.)"

When he wrote these words, Paul had no intention of experiencing this phase change anytime soon himself. As his 60th year approached, a colleague and former student wrote him to suggest organizing a scientific meeting in his honour. Paul's reply was dismissive: "Isn't that a bit premature? … I'm hardly moribund!" After further prodding he reluctantly supposed he might be ready to reconsider when he reached 70. That chance never came. A few days after his 66th birthday, during the early morning hours of September 16, 2015, Paul died peacefully in his sleep of a heart attack related to his ongoing medical condition. A Dr. Who fan until the end, his ashes rest in the Gabriola community cemetery beneath a gravestone reading "The TARDIS will take you now."

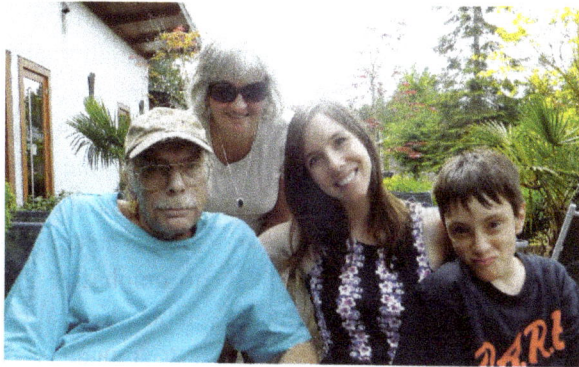

Fig. A.17: Paul (left) at home on Gabriola Island with life partner Pat, daughter Emily and son Sterling (2015)

A.11. Life and Legacy

Paul Wesson's life was as multidimensional as his science. His partner Pat recalled how his humble origins stamped him for life: he drove only used cars, did all his own home repairs, and would not hear of flying first-class even when he had the airline points to pay for it. His musical tastes were eclectic, ranging from Bob Dylan and the Kinks to Gustav Mahler and Olivier Messiaen. He admired the paintings of J.M.W. Turner and delighted in British comedy (especially Spike Milligan and Monty Python). He was devoted to vintage works of crime and science fiction, primarily from the 1950s and 1960s, and possessed complete collections of authors such as J.G. Ballard, Erle Stanley Gardner and Georges Simenon. His friends came from all walks of life. (The author of this biography first met him at a university pub, where the subject that drew us together was not physics but motorcycles.) He was a professed atheist, but a curious one with an open mind and a deep respect for the Quaker faith in particular.

As a scientist, Paul Wesson's work was marked by three characteristics. First, he was relentlessly hardworking and prolific. Second, like his heroes Eddington and Hoyle, he showed a consistent willingness to question accepted wisdom. As with those men at their best, this was no mere reflexive reaction to established authority, but rather an essentially creative activity, supported by extensive research. He was quite capable of letting go of a cherished idea, as he showed early on by abandoning his youthful infatuation with alternatives to continental drift. Third, though he was a quintessential theorist, he was instinctively uncomfortable with unfettered speculation. A typical inscription in the margins of one of his books on the string landscape reads: "After reading all of the wobble in this book, I feel redirected towards concrete, equation-based physics." To those who raised metaphysical issues with extra dimensions, he was most likely to emphasize the need for more exact mathematical solutions with acceptable physical properties. Above all, he took pains throughout his career to remain personally involved with experiment and observation. He passed this attitude on to his students as well, encouraging them to look up at the stars, but reminding them always to keep one foot on the ground.

Paul Wesson was among the fortunate few who stumble upon a big idea, and are able to devote much of their life to it. Time will tell whether there is truth to the idea that our four-dimensional world can be understood in terms of general relativity without sources in higher dimensions. But there is no doubt that, for Paul, this was more than a theory. He lived in the same troubled, ever-changing universe that the rest of us do, but for him it was something of an illusion, hiding an underlying reality both simpler and more austere. It is that deeper reality that Paul looked to describe with mathematics, to test with experiment, and ultimately to experience in death. His integrity as a scientist, colleague and friend will long inspire those of us he left behind.

Acknowledgments

Thanks go to the many people who shared their reminiscences and photographs, including Ron Adler, Allan Barros, Aurel Bejancu, Mauricio Bellini, Andrew Billyard, Stu Bowyer, Bernard Carr, Sujit Chatterjee, Martin Clutton-Brock, Alan Coley, Fred Cooperstock, Farhad Darabi, Walt Duley, Francis Everitt, Takao Fukui, Eduardo Guendelman, Paul Halpern, Dick Henriksen, Mark Israelit, Dimitri Kalligas, Tomas Liko, Bahram Mashhoon, Tom Mongan, Vesselin Petkov, Jaime Ponce de Leon, John Rather, Bill Sajko, Sanjeev Seahra, Jeff Secker, Alex Silbergleit, Rolf Stabell, Bob Stallman, Reza Tavakol, Alan Warrington, and Emily

Wesson. Special thanks are due to Jon Perry for technical help, and above all to Pat Lapcevic, who shared freely of her time and energy, supplying me with many critical memories, names and documents, and providing all the photographs used here (except where otherwise noted). Without her encouragement this book would not have been possible.

References

[1] Wesson, P.S. (1970). The Position Against Continental Drift, *Q. J. R. Astr. Soc.*, 11, pp. 312-340.

[2] Wesson, P.S. (1971). Convection and the Constant Q-Mechanism, *Nature*, 232, pp. 251-252.

[3] Wesson, P.S. (1972). Objections to Continental Drift and Plate Tectonics, *J. Geol.*, 80, pp. 185-197.

[4] Wesson, P.S. (1972). *Geological Report of the 1972 Cambridge Hindu Kush Expedition*, Dept. Geodesy and Geophysics, Cambridge University, 14 pp.

[5] Wesson, P.S. (1972). Mantle Creep: Elasticoviscous Versus Modified Lomnitz Law, and Problems of "The New Global Tectonics," *Am. Ass. Pet. Geol. Bull.*, 56, pp. 2127-2149.

[6] Wesson, P.S. (1973). Accretion and Electrostatic Interaction of Interstellar Dust Grains: Interstellar Grit, *Astrophys. Sp. Sci.*, 23, pp. 227-255.

[7] Wesson, P.S. (1973). The Implications for Geophysics of Modern Cosmologies in which *G* is Variable, *Q. J. R. Astr. Soc.*, 14, pp. 9-64.

[8] Wesson, P.S. (1974). Density Perturbations Set Up by a Rotating Galactic Dipole, *Astrophys. Sp. Sci.*, 26, pp. 189-197.

[9] Wesson, P.S. (1974). Problems of Plate Tectonics and Continental Drift, *Am. Ass. Pet. Geol. Memoir*, 23, pp. 146-154.

[10] Wesson, P.S. (1974). Vortex Flow of Matter Around a Singularity and a Galactic Hypothesis of Jeans, *Astrophys. Sp. Sci.*, 28, pp. 289-302.

[11] Wesson, P.S. (1974). Metric Conditions for Clusters in Hierarchical Cosmologies, *Astrophys. Sp. Sci.*, 30, pp. 71-94.

[12] Wesson, P.S. (1974). Expanding Clusters of Galaxies as Components of a Relativistic Hierarchical Cosmology, *Astrophys. Sp. Sci.*, 30, pp. 95-114.

[13] Wesson, P.S. (1974). Cosmological Deductions from the Relativistic Stability of Clusters of Galaxies, *Astrophys. Sp. Sci.*, 31, pp. 333-339.

[14] Wesson, P.S. (1974). Modified Lomnitz vs. Elasticoviscous Controversy, *Am. Ass. Pet. Geol. Memoir*, 23, pp. 448-462.

[15] Wesson, P.S. (1974). Application of a Non-Static Spherically-Symmetric Metric to Clusters of Galaxies, *Astrophys. Sp. Sci.*, 31, pp. 341-347.

[16] Wesson, P.S. (1974). A Synthesis of Our Present Knowledge of Interstellar Dust, *Sp. Sci. Rev.*, 15, pp. 469-482.

[17] Wesson, P.S. (1975). Relativistic Hierarchical Cosmology I: Derivation of a Metric and Dynamical Equations, *Astrophys. Sp. Sci.*, 32, pp. 273-284.

[18] Wesson, P.S. (1975). Relativistic Hierarchical Cosmology II: Some Classes of Model Universes, *Astrophys. Sp. Sci.*, 32, pp. 305-314.

[19] Wesson, P.S. (1975). Relativistic Hierarchical Cosmology III: Comparison with Observational Data, *Astrophys. Sp. Sci.*, 32, pp. 315-330.

[20] Wesson, P.S. (1975). Gravity and the Earth's Rotation, in Rosenberg, G.D. and Runcorn, S.K (eds.), *Growth Rhythms and the History of the Earth's Rotation* (Wiley, New York), pp. 353-374.

[21] Wesson, P.S. (1975). The Interrelationship between Cosmic Dust and the Microwave Background, *Astrophys. Sp. Sci.*, 36, pp. 363-382.

[22] Wesson, P.S. (1975). Discrete Source Counts and the Rotation of the Local Universe in Hierarchical Cosmology, *Astrophys. Sp. Sci.*, 37, pp. 235-256.

[23] Wesson, P.S. (1975). Tests of Randomness in Astronomical Objects. I: Along Single Great Circles, *Australian J. Phys.*, 28, pp. 453-461.

[24] Wesson, P.S. (1975). Tests of Randomness in Astronomical Objects. II: Along Two Orthogonal Great Circles, *Australian J. Phys.*, 28, pp. 463-473.

[25] Wesson, P.S. (1976). Covariance Function Analysis and the Clustering of Galaxies, *Astrophys. Sp. Sci.*, 40, pp. 325-349.

[26] Wesson, P.S. (1976). Expanding and Superdense Astrophysical Systems in General Relativistic Hierarchial Cosmology, *Astrophys. Sp. Sci.*, 42, pp. 285-330.

[27] Wesson, P.S. and Lermann, A. (1976). Limits on Cosmic Dust Densities by Observations of the Microwave Background, *Astron. Astrophys.*, 53, pp. 383-388.

[28] Wesson, P.S. (1976). Symmetry in Hierarchical Cosmology, *Astrophys. Sp. Sci.*, 42 pp. 477-481.

[29] Wesson, P.S. and Lermann, A. (1977). Hydromagnetic Breakup of Bridges and Jets into Aligned Objects, *Astrophys. Sp. Sci.*, 46, pp. 51-60.

[30] Wesson, P.S. (1977). Microwave Emission from Dust and Hidden Matter in the Coma Cluster, *Astron. Astrophys.*, 61, pp. 177-180.

[31] Wesson, P.S. (1977). Cosmic Rays and Ancient Planetary Magnetic Fields, *Astrophys. Sp. Sci.*, 46, pp. 321-326.

[32] Wesson, P.S. and Lermann, A. (1977). Substructure in Clusters of Galaxies: A Clue to the 'Missing' Mass? *Astrophys. Sp. Sci.,* 46, pp. 327-334.

[33] Wesson, P.S. (1977). The Case of the Missing Cosmic Dust, *New Scientist*, 73 (27 Jan. 1977), pp. 207-209.

[34] Wesson, P.S. and Lermann, A. (1977). A Hypothesis to Explain the Virial Discrepancy in Clusters of Galaxies, *Astrophys. Sp. Sci.*, 46, pp. 335-339.

[35] Wesson, P.S. (1977). Does the Gravitational Constant Vary With Time? *Astronomisk Tidsskrift*, 10, pp. 162-163 (in Norwegian).

[36] Wesson, P.S., Lermann, A. and Goodson, R.E. (1977). Numerical Experiments on the Effect of Massive Central Systems on the Virial Binding Energy of Clusters, *Astrophys. Sp. Sci.*, 48, pp. 357-363.

[37] Wesson, P.S. (1978). *Cosmology and Geophysics* (Oxford University Press/Hilger, New York), 240 pp.

[38] Wesson, P.S. and Lermann, A. (1978). The Formation of Iron, Stone and Mixed Planetesimals in the Early Solar System. *Icarus*, 33, pp. 74-88.

[39] Wesson, P.S. and Lermann, A. (1978). The Configuration of a Thin, Rotating Self-Similar Disk in General Relativity, *Astrophys. Sp. Sci.*, 57, pp. 203-217; *Erratum* (1979), *ibid*, 60, p. 242.

[40] Wesson, P.S. (1978). A New Look at the Cosmological Principle, *Astron. Astrophys.*, 68, pp. 131-137.

[41] Wesson, P.S. (1978). General Relativistic Hierarchical Cosmology: An Exact Model, *Astrophys. Sp. Sci.*, 54, pp. 489-495.

[42] Wesson, P.S. (1978). The Number Density Excess of QSOs and Initially Inhomogeneous Cosmologies, *Astrophys. Lett.*, 19, pp. 121-125.

[43] Wesson, P.S. (1978). Anticorrelation and Galactic Haloes, *Astrophys. Lett.*, 19, pp. 127-131.

[44] Wesson, P.S. (1978). Dynamic Measurement of Matter Creation and Earth Expansion, *Nature*, 273, p. 572.

[45] Wesson, P.S. (1978). An Exact Solution to Einstein's Equations with a Stiff Equation of State, *J. Math. Phys.*, 19, pp. 2283-2284.

[46] Henriksen, R.N. and Wesson, P.S. (1978). Self-Similar Space-Times. I: Three Solutions, *Astrophys. Sp. Sci.*, 53, pp. 429-444.

[47] Henriksen, R.N. and Wesson, P.S. (1978). Self-Similar Space-Times. II: Perturbation Scheme, *Astrophys. Sp. Sci.*, 53, pp. 445-457.

[48] Wesson, P.S. (1978). Dust and the Formation of Planets with Iron Cores in the Early Solar System, *Astron. Astrophys.*, 69, pp. 125-128.

[49] Wesson, P.S. (1979). The Cosmological Principle, *Astronomy*, 7 (10), pp. 66-70.

[50] Wesson, P.S. (1979). Observable Relations in an Inhomogeneous Self-Similar Cosmology, *Astrophys. J.*, 228, pp. 647-663.

[51] Wesson, P.S. (1979). The Formation of Planets and Satellites from Self-Similar Disks, *Astron. Astrophys.*, 76, pp. 200-207.

[52] Wesson, P.S. (1979). Self-Similarity and the Angular Momenta of Astronomical Systems: A Basic Rule in Astronomy, *Astron. Astrophys.*, 80, pp. 296-300.

[53] Wesson, P.S. (1980). *Gravity, Particles, and Astrophysics* (Reidel, Dordrecht), 188 pp.

[54] Wesson, P.S. (1980). Theories for the Origin of the Solar System, *Astronomisk Tidsskrift*, 13, pp. 58-61 (in Norwegian).

[55] Wesson, P.S. (1980). Does the Binding Energy of Binaries Masquerade as Missing Mass? *Astron. Astrophys.*, 90, pp. 1-7.

[56] Wesson, P.S. (1980). Does Gravity Change with Time? *Physics Today*, 33 (7), pp. 32-37.

[57] Stabell, R. and Wesson, P.S. (1980). Background Light from Galaxies as a Cosmological Probe, *Astrophys. J.*, 242, pp. 443-447.

[58] Wesson, P.S. (1980). A Review of Likely Theories of the Origin of the Solar System, *Solar System Today*, 2, pp. 43-47.

[59] Wesson, P.S. (1980). The Application of Dimensional Analysis to Cosmology (or: How to Make Cosmology Simple by Using Dimensional Conspiracy), *Sp. Sci. Rev.*, 27, pp. 109-153.

[60] Wesson, P.S. (1981). The Significance of Electrostatic Accretion Forces between Dust Grains, *Moon and Planets*, 24, pp. 339-343.

[61] Wesson, P.S. (1981). Clue to the Unification of Gravitation and Particle Physics, *Phys. Rev.*, D23, pp. 1730-1734.

[62] Wesson, P.S. (1981). Is the Universe Spinning? *Astronomy*, 9 (1), pp. 67-71.

[63] Wesson, P.S. (1981). Improved Standard Cosmology, *Phys. Rev.*, D23, pp. 2137-2141.

[64] Wesson, P.S. (1981). How Homogeneous is the Universe? *Astrophys. Lett.*, 21, pp. 97-99.

[65] Wesson, P.S. (1981). Scale-Invariant Gravity: A Reformulation and an Astrophysical Test, *Mon. Not. R. Astr. Soc.*, 197, pp. 157-165.

[66] Wesson, P.S. and Goodson, R.E. (1981). New Pathways in Gravitational Research, *Observatory*, 101, pp. 105-108.

[67] Wesson, P.S. (1981). Scale-Invariant Gravity: A Simple Formulation, *Astron. Astrophys.*, 102, pp. 45-52.

[68] Wesson, P.S. (1982). Gravitational Interactions and the Origin of the Angular Momenta of Galaxies, *Vistas Astron.*, 25, pp. 411-418.

[69] Wesson, P.S. (1982). Implications of Gravitational Interactions for the Angular Momenta of Galaxies, *Astrophys. Lett.*, 23, pp. 1-5.

[70] Wesson, P.S. (1982). Improved Standard Cosmology: Comparison with Observation, *Astrophys. J.*, 259, pp. 20-24.

[71] Wesson, P.S. (1982). Galaxies in Clusters: Alignments, Formation from Pancakes, and Tidal Forces, *Vistas Astron.*, 26, pp. 225-241.

[72] Wesson, P.S. (1983). A New Approach to Scale-Invariant Gravity, *Astron. Astrophys.*, 119, pp. 145-152.

[73] Wesson, P.S. (1983). Inflationary Universe Models with a Conformal Killing Vector, in Small, R.D. (ed.), *Proc. Atlantic Math. Days*, Dept. Math. and Statistics, University of New Brunswick, Canada, pp. 101-118.

[74] Wesson, P.S. (1983). A Note on the Formation of Clusters of Galaxies, *Astrophys. J.*, 268, pp. 561-564.

[75] Wesson, P.S. (1983). Clarification of the Angular Momentum/Mass Relation ($J = pM^2$) for Astronomical Objects, *Astron. Astrophys.*, 119, pp. 313-314.

[76] Henriksen, R.N., Emslie, A.G. and Wesson, P.S. (1983). Space-Times with Constant Vacuum Energy Density and a Conformal Killing Vector, *Phys. Rev.*, D27, pp. 1219-1227.

[77] Wesson, P.S. (1984). Tidal Torques and the Orientations of Galaxies in Clusters, *Astron. Astrophys.*, 138, pp. 253-257.

[78] Wesson, P.S. (1984). An Embedding for General Relativity with Variable Rest Mass, *Gen. Rel. and Grav.*, 16, pp. 193-203.

[79] Wesson, P.S. (1984). Protostars and the Origin of the Angular Momentum of the Solar System, *Earth, Moon and Planets*, 30, pp. 275-280.

[80] Wesson, P.S. (1984). A Cosmological Solution of Einstein's Equations, *J. Math. Phys.*, 25, pp. 3297-3298.

[81] Wesson, P.S. (1984). Dimensional Fundamental Constants and Their Application in Astronomy, *Sp. Sci. Rev.*, 39, pp. 153-161.

[82] Wesson, P.S. (1985). Relations for a Kaluza-Klein Cosmology with Variable Rest Mass, *Astron. Astrophys.*, 143, pp. 233-234.

[83] Wesson, P.S. and Stabell, R. (1985). Has the Earth Grown by 20% in 200 Million Years? *Forsknings Nytt*, 30 (4), pp. 21-24 (in Norwegian).

[84] Wesson, P.S. (1985). Tidal Torques and the Spins of Galaxies Formed in the Adiabatic or Pancake Theory, *Astron. Astrophys.*, 151, pp.105-107.

[85] Wesson, P.S. (1985). Avoiding the Big Bang: Matter Production from Minkowski Space and the Early Universe, *Astron. Astrophys.*, 151, pp.276-278.

[86] Wesson, P.S. (1986). The Extra-Galactic Background Light: A Modern Version of Olbers' Paradox, *Sp. Sci. Rev.*, 44, pp. 169-176.

[87] Wesson, P.S. (1986). Astronomy: What's Happening in Canada, *Phys. Bull.*, 37 (7), pp. 284-285.

[88] Wesson, P.S. (1986). Astrophysical Data and Cosmological Solutions of a Kaluza-Klein Theory of Gravity, *Astron. Astrophys.*, 166, pp. 1-3.

[89] Wesson, P.S. (1986). New Ideas in Cosmology and New Solutions in General Relativity, *Vistas Astron.*, 29, pp. 281-289.

[90] Wesson, P.S. (1986). Comments on a Class of Similarity Solutions of Einstein's Equations Relevant to the Early Universe, *Phys. Rev.*, D34, pp. 3925-3926.

[91] Wesson, P.S. (1987). Gravitational Theory and the Clustering of Galaxies, in Stoeger, W.R. (ed.), *Theory and Observational Limits in Cosmology* (Vatican Observatory, Rome), pp. 547-558.

[92] Wesson, P.S., 1987. Problems with the Way in Which Cosmology is Done, in Stoeger, W.R. (ed.), *Theory and Observational Limits in Cosmology* (Vatican Observatory, Rome), pp. 559-563.

[93] Wesson, P.S., Valle, K. and Stabell, R. (1987). The Extragalactic Background Light and a Definitive Resolution of Olbers' Paradox, *Astrophys. J.*, 317, pp. 601-606.

[94] Wesson, P.S. (1988). Cosmology without the Big Bang, *Analog*, 108 (1), pp. 36-43.

[95] Wesson, P.S. (1988). Comments on a Possible Change with Cosmological Time in the Rest Masses of Particles, *Astron. Astrophys.*, 189, pp. 4-6.

[96] Stabell, R. and Wesson, P.S. (1988). Why the Night Sky is Dark, *Populaervitenskapelige Magasinet* 1, p. 58 (in Norwegian).

[97] Wesson, P.S. (1988). The Radiation Field Due to Galaxies in Friedmann-Robertson-Walker Models, in Coley, A., Dyer, C. and Tupper, B. (eds.), *Proc. 2nd Canadian*

Conference on General Relativity and Relativistic Astrophysics (World Scientific, Singapore), pp. 105-116.

[98] Gribbin, J. and Wesson, P.S. (1988). The Fifth Dimension of Mass, *New Scientist*, 119 (22 Sept. 1988), pp. 56-59.

[99] Wesson, P.S. and Ponce de Leon, J. (1988). Astrophysical and Cosmological Solutions in General Relativity: A Comment on One Form of the Spherically-Symmetric Perfect-Fluid Field Equations, *Astron. Astrophys., 206*, pp. 7-8.

[100] Wesson, P.S. (1989). A Class of Solutions in General Relativity of Interest for Cosmology and Astrophysics, *Astrophys. J., 336*, pp. 58-60.

[101] Wesson, P.S. (1989). Olbers' Paradox in Astronomy: History, Nature and Resolution, *Sci. Prog., 73*, pp. 133-146.

[102] Wesson, P.S. and Stabell, R. (1989). The Dimensions of Modern Cosmology, *Astronomisk Tidsskrift*, 22, pp. 28-31 (in Norwegian).

[103] Wesson, P.S. (1989). Olbers' Paradox Solved At Last, *Sky and Tel.*, 77, pp. 594-597.

[104] Wesson, P.S. and Ponce de Leon, J. (1989). Cosmological Solution of Einstein's Equations with Uniform Density and Nonuniform Pressure, *Phys. Rev.*, D39, pp. 420-422.

[105] Wesson, P.S. (1989). The Real Reason the Night Sky is Dark: Correcting a Myth in Astronomy Teaching, *J. British Astron. Ass.*, 99, pp. 10-13.

[106] Wesson, P.S. (1990). Variable-Gravity Theory: A New Version of an Old Idea, in Coley, A., Cooperstock, F. and Tupper, B. (eds.), *Proc. 3rd Canadian Conference on General Relativity and Relativistic Astrophysics* (World Scientific, Singapore), pp. 9-13.

[107] Wesson, P.S. (1990). Cosmology, Extraterrestrial Intelligence, and a Resolution of the Fermi-Hart Paradox, *Q. J. R. Astr. Soc.*, 31, pp. 161-170.

[108] Wesson, P.S. (1990). Clarification of an Extended Theory of Gravity and a Reply to Gron and Soleng, *Gen. Rel. Grav.*, 22, pp. 707-713.

[109] Wesson, P.S. (1990). Inhomogeneous Solutions in Cosmology, *Can. J. Phys.*, 68, pp. 824-826.

[110] Wesson, P.S. (1990). The Fine-Structure Constant and Eddington, *Physics Today*, 43 (12), p. 15.

[111] Mann, R.B. and Wesson, P.S. (1991). *Gravitation: A Banff Summer Institute* (World Scientific, Singapore), 650 pp.

[112] Wesson, P.S. (1991). Olbers' Paradox and the Spectral Intensity of the Extragalactic Background Light, *Astrophys. J.*, 367, pp. 399-406.

[113] Wesson, P.S. (1991). Cosmological Source of Vacuum Electromagnetic Zero-Point Energy, *Phys. Rev.*, A44, pp. 3379-3381.

[114] Wesson, P.S. (1991). Testing the Equivalence Principle in the Rocky Mountains, in Mann, R.B. and Wesson, P.S. (eds.), *Gravitation: A Banff Summer Institute* (World Scientific, Singapore), pp. 638-642.

[115] Wesson, P.S. (1991). Cosmological Constraints on the Zero-Point Electromagnetic Field, *Astrophys. J.*, 378, pp. 466-470.

[116] Wesson, P.S. (1992). The Properties of Matter in Kaluza-Klein Cosmology, *Mod. Phys. Lett.*, A7, pp. 921-926.

[117] Wesson, P.S. (1992). Constants and Cosmology: The Nature and Origin of Fundamental Constants in Astrophysics and Particle Physics, *Sp. Sci. Rev.*, 59, pp. 365-406.

[118] Wesson, P.S. (1992). The Effective Properties of Matter of Kaluza-Klein Solitons, *Phys. Lett.*, B276, pp. 299-302.

[119] Gribbin, J. and Wesson, P.S. (1992). Fickle Constants of Physics, *New Scientist*, 135 (4 July 1992), pp. 30-33.

[120] Wesson, P.S. (1992). A Physical Interpretation of Kaluza-Klein Cosmology, *Astrophys. J.*, 394, pp. 19-24.

[121] Kalligas, D., Wesson, P.S. and Everitt, C.W.F. (1992). Flat FRW Models with Variable G and Λ. *Gen. Rel. Grav.*, 24, pp. 351-357.

[122] Wesson, P.S. and Lim. P. (1992). Astrophysical Implications of a Solution in Kaluza-Klein Gravity, *Astron. Astrophys.*, 261, pp. 373-376.

[123] Overduin, J.M. and Wesson, P.S. (1992). Background Radiation from Dark Matter, in Kunstatter, G., Vincent, D.E. and Williams, J.G. (eds.), *Proc. 4th Canadian Conference on General Relativity and Relativistic Astrophysics* (World Scientific, Singapore), pp. 332-335.

[124] Lim, P. and Wesson, P.S. (1992). The Perihelion Problem in Kaluza-Klein Gravity, *Astrophys. J.*, 397, pp. L91-94.

[125] Wesson, P.S. (1992). Cosmology and Zero-Point Fields, *Phys. Essays*, 5, pp. 561-566.

[126] Liu, H. and Wesson, P.S. (1992). Exact Solutions of General Relativity Derived from 5D "Black Hole" Solutions of Kaluza-Klein Theory, *J. Math. Phys.*, 33, pp. 3888-3891.

[127] Overduin, J.M. and Wesson, P.S. (1992). Constraints on Dark Matter from Intergalactic Radiation, *Vistas Astron.*, 35, pp. 439-478.

[128] Wesson, P.S. and Ponce de Leon, J. (1992). Kaluza-Klein Equations, Einstein's Equations and an Effective Energy-Momentum Tensor, *J. Math. Phys.*, 33, pp. 3883-3887.

[129] Overduin, J.M., Wesson, P.S. and Bowyer, S. (1993). Constraints on Vacuum Decay from the Microwave Background, *Astrophys. J.*, 404, pp. 1-7.

[130] Wesson, P.S., Liu, H. and Lim, P. (1993). Time-Dependent Kaluza-Klein Solitons and Their Effective Properties of Matter, *Phys. Lett.*, B298, pp. 69-72.

[131] Overduin, J.M., Wesson, P.S. and Bowyer, S. (1993). Constraints on Decaying Neutrinos from the Far-Ultraviolet Extragalactic Background Light, *Astrophys. J.*, 404, pp. 460-467.

[132] Ponce de Leon, J. and Wesson, P.S. (1993). Exact Solutions and the Effective Equation of State in Kaluza-Klein Theory, *J. Math. Phys.*, 34, pp. 4080-4092.

[133] Overduin, J.M. and Wesson, P.S. (1993). Constraints on Axions from the Extragalactic Background Light, *Astrophys. J.*, 414, pp. 449-453.

[134] Wesson, P.S., Ponce de Leon, J., Lim, P. and Liu, H. (1993). Physical Properties of Matter Derived from Geometry of Kaluza-Klein Theory, *Int. J. Mod. Phys.*, D2, pp. 163-170.

[135] Liu, H., Wesson, P.S. and Ponce de Leon, J. (1993). Time-Dependent Kaluza-Klein Soliton Solutions, *J. Math. Phys.*, 34, pp. 4070-4079.

[136] Wesson, P.S. (1994). A New Dark Matter Candidate: Kaluza-Klein Solitons, *Astrophys. J.*, 420, pp. L49-52.

[137] Liu, H. and Wesson, P.S. (1994). Cosmological Solutions with Radiation in Kaluza-Klein Gravity, in Mann, R.B. and McLenaghan, R.G. (eds.), *Proc. 5th Canadian Conference on General Relativity and Relativistic Astrophysics* (World Scientific, Singapore), pp. 433-437.

[138] Liu, H., Wesson, P.S. (1994). Matter Properties of a Kaluza-Klein Cosmological Soliton Solution, in Mann, R.B. and McLenaghan, R.G. (eds.), *Proc. 5th Canadian Conference on General Relativity and Relativistic Astrophysics* (World Scientific, Singapore), pp. 438-442.

[139] Wesson, P.S. and Ponce de Leon, J. (1994). Kaluza-Klein Theory and Machian Cosmology, *Gen. Rel. Grav.*, 26, pp. 555-566.

[140] Liu, H. and Wesson, P.S. (1994). Cosmological Solutions and Their Effective Properties of Matter in Kaluza-Klein Theory, *Int. J. Mod. Phys.*, D3, pp. 627-637.

[141] Wesson, P.S. and Ponce de Leon, J. (1994). The Physical Properties of Kaluza-Klein Solitons, *Class. Quant. Grav.*, 11, pp. 1341-1351.

[142] Secker, J., Lepock, J. and Wesson, P.S. (1994). Damage Due to Ultraviolet and Ionizing Radiation during the Ejection of Shielded Micro-Organisms from the Vicinity of Solar-Mass Main Sequence and Red Giant Stars, *Astrophys. Sp. Sci.*, 219, pp. 1-28.

[143] Wesson, P.S. (1994). An Embedding for the Big Bang, *Astrophys. J.*, 436, pp. 547-550.

[144] Mashhoon, B., Liu, H. and Wesson, P.S. (1994). Particle Masses and the Cosmological Constant in Kaluza-Klein Theory, *Phys. Lett.*, B331, pp. 305-312.

[145] Wesson, P.S. and Ponce de Leon, J. (1995). The Equation of Motion in Kaluza-Klein Cosmology and Its Implications for Astrophysics, *Astron. Astrophys.*, 294, pp. 1-7.

[146] Kalligas, D., Wesson, P.S. and Everitt, C.W.F. (1995). The Classical Tests in Kaluza-Klein Gravity, *Astrophys. J.*, 439, pp. 548-557.

[147] Wesson, P.S. and Liu, H. (1995). Fully Covariant Cosmology and its Astrophysical Implications, *Astrophys. J.*, 440, pp. 1-4.

[148] Wesson, P.S. (1995). Consequences of Covariance in Kaluza-Klein Theory, *Mod. Phys. Lett.*, A10, pp. 15-24.

[149] Kalligas, D., Wesson, P.S. and Everitt, C.W.F. (1995). Bianchi Type I Cosmological Models with Variable G and Λ: A Comment, *Gen. Rel. Grav.*, 27, pp. 645-650.

[150] Billyard, A., Wesson, P.S. and Kalligas, D., 1995. Physical Aspects of Solitons in (4+1) Gravity, *Int. J. Mod. Phys.*, D4, pp. 639-659.

[151] Lim, P., Overduin, J.M. and Wesson, P.S. (1995). Light Deflection in Kaluza-Klein Gravity, *J. Math. Phys.*, 36, pp. 6907-6914.

[152] Billyard, A. and Wesson, P.S. (1996). Waves in Five-Dimensional Relativity Theory, *Gen. Rel. Grav.*, 28, pp. 129-136.

[153] Billyard, A. and Wesson, P.S. (1996). Class of Exact Solutions in 5D Gravity and Its Physical Properties, *Phys. Rev.*, D53, pp. 731-737.

[154] Wesson, P.S., Ponce de Leon, J., Liu, H., Mashhoon, B., Kalligas, D., Everitt, C.W.F., Billyard, A., Lim, P. and Overduin, J.M. (1996). A Theory of Space, Time and Matter, *Int. J. Mod. Phys.*, A11, pp. 3247-3255.

[155] Liu, H. and Wesson, P.S. (1996). The Motion of a Spinning Object in a Higher-Dimensional Spacetime, *Class. Quant. Grav.*, 13, pp. 2311-2318.

[156] Liu, H. and Wesson, P.S. (1996). A Class of Kaluza-Klein Soliton Solutions, *Phys. Lett.*, B381, pp. 420-422.

[157] Secker, J., Wesson, P.S. and Lepock, J.R. (1996). Astrophysical and Biological Constraints on Radiopanspermia, *J. R. Astr. Soc. Canada*, 90, pp. 184-192.

[158] Wesson, P.S. (1996). Possible Violations of the Equivalence Principle in Kaluza-Klein Theories, in Reinhard, R. (ed.), *Proc. Symposium on Satellite Test of the Equivalence Principle* (European Space Agency, WPP-115), pp. 566-572.

[159] Billyard, A. and Wesson, P.S. (1996). Kaluza-Klein Solitons: New Singularities and Origin Definitions, in Jantzen, R.T., Keiser, G.M. and Ruffini, R. (eds.), *Proc. 7th. Marcel Grossmann Meeting on General Relativity* (World Scientific, Singapore), pp. 317-318.

[160] Kalligas, D., Wesson, P.S. and Everitt, C.W.F. (1996). The Classical Tests in (4+1)-Gravity, in Jantzen, R.T., Keiser, G.M. and Ruffini, R. (eds.), *Proc. 7th. Marcel Grossmann Meeting on General Relativity* (World Scientific, Singapore), pp. 330-332.

[161] Mashhoon, B., Liu, H. and Wesson, P.S. (1996). Space-Time-Matter, in Jantzen, R.T., Keiser, G.M. and Ruffini, R. (eds.), *Proc. 7th. Marcel Grossmann Meeting on General Relativity* (World Scientific, Singapore), pp. 333-335.

[162] Overduin, J.M., Lim, P. and Wesson, P.S., 1996. Lensing in (4+1) Gravity, in Jantzen, R.T., Keiser, G.M. and Ruffini, R. (eds.), *Proc. 7th. Marcel Grossmann Meeting on General Relativity* (World Scientific, Singapore), pp. 1151-1152.

[163] Wesson, P.S. (1996). Geometry and Matter in (4+1) Gravity, in Jantzen, R.T., Keiser, G.M. and Ruffini, R. (eds.), *Proc. 7th. Marcel Grossmann Meeting on General Relativity* (World Scientific, Singapore), pp. 1156-1158.

[164] Billyard, A. and Wesson, P.S. (1997). A Wave-Like Solution in Kaluza-Klein Theory, in Braham, S.P., Gegenberg, J.D. and McKellar, R.J. (eds.), *Proc. 6th Canadian Conference on General Relativity and Relativistic Astrophysics* (Am. Math. Soc., Providence, Rhode Island), pp. 161-165.

[165] Kalligas, D. and Wesson, P.S. (1997). Light Repulsion in Higher-Dimensional Relativity, in Braham, S.P., Gegenberg, J.D. and McKellar, R.J. (eds.), *Proc. 6th*

Canadian Conference on General Relativity and Relativistic Astrophysics (Am. Math. Soc., Providence, Rhode Island), pp. 259-262.

[166] Ponce de Leon, J. and Wesson, P.S. (1997). The Equation of Motion in Higher-Dimensional Gravity, in Braham, S.P., Gegenberg, J.D. and McKellar, R.J. (eds.), *Proc. 6th Canadian Conference on General Relativity and Relativistic Astrophysics* (Am. Math. Soc., Providence, Rhode Island), pp. 325-329.

[167] Overduin, J.M. and Wesson, P.S. (1997). Kaluza-Klein Gravity, *Physics Reports,* 283, pp. 303-378.

[168] Overduin, J.M. and Wesson, P.S. (1997). Background Radiation Constraints on Supersymmetric Weakly Interacting Particles, *Astrophys. J.,* 480, pp. 470-480.

[169] Wesson, P.S., Secker, J. and Lepock, J.R. (1997). Panspermia Revisited: Astrophysical and Biological Constraints, in Cosmovici, C.B., Bowyer, S. and Wertheimer, D. (eds.), *Astronomical and Biochemical Origins and the Search for Life in the Universe* (Ed. Compositori, Bologna), pp. 539-544.

[170] Wesson, P.S. and Liu, H. (1997). Mass and Charge from Higher-Dimensional Geometry, *Int. J. Theor. Phys.,* 36, pp. 1865-1879.

[171] Overduin, J.M. and Wesson, P.S. (1997). Decaying Neutrinos and the Extragalactic Background Light, *Astrophys. J.,* 483, pp. 77-86.

[172] Wesson, P.S. (1997). Comments on the Cosmological Constant Problem and how it Can be Solved, *Int. J. Mod. Phys.,* D6, pp. 643-648.

[173] Chatterjee, S., Wesson, P.S. and Billyard, A. (1997). Cylindrically-Symmetric Spacetimes in 5D and Their Effective Properties of Matter, *Il Nuovo Cim.,* B112, pp. 1667-1673.

[174] Wesson, P.S., Mashhoon, B. and Liu, H. (1997). The (Im)possibility of Detecting a Fifth Dimension, *Mod. Phys. Lett.,* A12, pp. 2309-2316.

[175] Wesson, P.S. (1997). Neutrinos, Axions and the Vacuum: Some Cosmological Constraints, in De Sabbata, V. and Singh, T. (eds.), *New Directions in Relativity and Cosmology* (Hadronic Press, Florida), pp. 209-214.

[176] Liu, H. and Wesson, P.S. (1997). The Physical Properties of Charged Five-Dimensional Black Holes, *Class. Quant. Grav.,* 14, pp. 1651-1663.

[177] Liu, H. and Wesson, P.S. (1998). A Class of Kaluza-Klein Solutions Curved in 4D and Flat in 5D, *Gen. Rel. Grav.,* 30, pp. 509-514.

[178] Wesson, P.S. and Liu, H. (1998). Shell-Like Solutions in Kaluza-Klein Theory, *Phys. Lett.,* B 432, pp. 266-270.

[179] Mashhoon, B., Wesson, P.S. and Liu, H. (1998). Dynamics in Kaluza-Klein Gravity and a Fifth Force, *Gen. Rel. Grav.,* 30, pp. 555-571.

[180] Liu, H. and Wesson, P.S (1998). Kaluza-Klein-Gordon Field Theory, *Int. J. Mod. Phys.,* D 5, pp. 737-747.

[181] Liu, H. and Wesson, P.S. (1998). Particle Masses in Kaluza-Klein-Gordon Theory and the Reality of Extra Dimensions, *Mod. Phys. Lett.,* A 13, pp. 2689-2694.

[182] Sajko, W.N., Wesson, P.S. and Liu, H. (1998). Gauge Conditions in Modern Kaluza-Klein Theory, *J. Math. Phys.,* 39, pp. 2193-2200.

[183] Overduin, J.M. and Wesson, P.S. (1998). Kaluza-Klein Cosmology with Noncompactified Extra Dimensions, in Rainer, M. and Schmidt, H.-J. (eds.), *Current Topics in Mathematical Cosmology* (World Scientific, Singapore), pp. 293-301.

[184] Wesson, P.S. (1999). *Space-Time-Matter* (World Scientific, Singapore), 209 pp; (2007, 2nd ed.), 254 pp.

[185] Agnese, G., Billyard, A., Liu, H. and Wesson, P.S. (1999). Possible Wormhole Solutions in (4+1) Gravity, *Gen. Rel. Grav.*, 31, pp. 527-535.

[186] Seahra, S.S., Overduin, J.M., Duley, W.W. and Wesson, P.S. (1999). A Critical Examination of Sciama's Heavy Neutrino Hypothesis, in Burgess, C.P. and Myers, R.C. (eds.), *Proc. 8th Canadian Conference on General Relativity and Relativistic Astrophysics,* (Am. Inst. Phys., New York), pp. 219-222.

[187] Wesson, P.S. and Sajko, W.N. (1999). Definitions of Particle Mass in Kaluza-Klein Theory, in Burgess, C.P. and Myers, R.C. (eds.), *Proc. 8th Canadian Conference on General Relativity and Relativistic Astrophysics,* (Am. Inst. Phys., New York), pp. 262-266.

[188] Sajko, W.N. and Wesson, P.S. (1999). Energy and Motion in Kaluza-Klein Gravity, in Burgess, C.P. and Myers, R.C. (eds.), *Proc. 8th Canadian Conference on General Relativity and Relativistic Astrophysics,* (Am. Inst. Phys., New York), pp. 267-271.

[189] Sajko, W.N., Wesson, P.S. and Liu, H. (1999). Waves and Particles in Kaluza-Klein Theory, *J. Math. Phys.*, 40, pp. 2364-2380.

[190] Overduin, J.M., Seahra, S.S., Duley, W.W. and Wesson, P.S. (1999). Could Intergalactic Dust Obscure a Neutrino Decay Signature? *Astron. Astrophys.*, 349, pp. 317-322.

[191] Wesson, P.S., Mashhoon, B., Liu, H. and Sajko, W.N. (1999). Fifth Force from Fifth Dimension, *Phys. Lett., B*, 456, pp. 34-37.

[192] Liu, H. and Wesson, P.S. (2000). On the Klein-Gordon Equation in Higher Dimensions: Are Particle Masses Variable? *Gen. Rel. Grav.*, 32, pp. 583-592.

[193] Wesson, P.S. (2000). On the Re-Emergence of Eddington's Philosophy of Science, *Observatory*, 120, pp. 59-62.

[194] Overduin, J.M. and Wesson, P.S., 2000. Observational Limits on Nonstandard Dark Matter Candidates, in Klapdor-Kleingrothaus, H.V. and Krivosheina, I.V. (eds.), *Beyond the Desert 1999* (Inst. Phys., Bristol), pp. 539-544.

[195] Sajko, W.N. and Wesson, P.S. (2000). The Energy of Asymptotically de Sitter Spacetimes in Kaluza-Klein Theory, *Gen. Rel. Grav.*, 32, pp. 1381-1388.

[196] Darabi, F., Sajko, W.N. and Wesson, P.S. (2000). Quantum Cosmology of 5D Non-Compactified Kaluza-Klein Theory, *Class. Quant. Grav.*, 17, pp. 4357-4364.

[197] Wesson, P.S., Liu, H. and Seahra, S.S. (2000). The Big Bang as a Higher-Dimensional Shock Wave, *Astron. Astrophys.*, 358, pp. 425-427.

[198] Wesson, P.S. (2000). *Zero-Point Fields, Gravitation and New Physics* (white paper, California Institute for Physics and Astrophysics, Palo Alto), available online at http://www.zpower.com/sp/documents/ZPEPaper_ZeroPointFieldsGravitationAnd NewPhysics.pdf (3/20/2018), 21 pp.

[199] Wesson, P.S. (2000). Utopia and Theoretical Physics, *Literary Review of Canada*, 8 (6), pp. 13-14.

[200] Wesson, P.S. and Liu, H. (2001). The Cosmological Constant Problem and Kaluza-Klein Theory, *Int. J. Mod. Phys.*, D10, pp. 905-912.

[201] Liu, H. and Wesson, P.S. (2001). Radiating Sources in Higher-Dimensional Gravity, *J. Math. Phys.*, 42, pp. 4963-4970.

[202] Wesson, P.S. (2001). Is Time Instantaneous? *Observatory*, 121, pp. 82-86.

[203] Sajko, W.N. and Wesson, P.S. (2001). The Energy of 5D Solitons, *Mod. Phys. Lett.*, A16, pp. 627-632.

[204] Wesson, P.S. (2001). Fundamental Unsolved Problems in Astrophysics, *Sp. Sci. Rev.*, 98, pp. 329-342.

[205] Seahra, S.S. and Wesson, P.S. (2001). Null Geodesics in Five-Dimensional Manifolds, *Gen. Rel. Grav.*, 33, pp. 1731-1752.

[206] Liu, H. and Wesson, P.S. (2001). Universe Models with a Variable Cosmological "Constant" and a "Big Bounce," *Astrophys. J.*, 562, pp. 1-6.

[207] Fukui, T., Seahra, S.S. and Wesson, P.S. (2001). Cosmological Implications of a Non-Separable 5D Solution of the Vacuum Einstein Field Equations, *J. Math. Phys.*, 42, pp. 5195-5201.

[208] Wesson, P.S. and Seahra, S.S. (2001). Images of the Big Bang, *Astrophys. J.*, 558, pp. L75-78.

[209] Wesson, P.S. (2002). *The Interstellar Undertakers* (Vantage Press, New York), 202 pp.

[210] Wesson, P.S., Seahra, S.S. and Liu, H. (2002). A Formal Approach to Machian General Relativity, *Int. J. Mod. Phys.*, D11, pp. 1347-1354.

[211] Cho, Y.M., Neupanee, I. and Wesson, P.S. (2002). No Ghost States of Gauss-Bonnet Interaction in Warped Backgrounds, *Nucl. Phys.*, B621, pp. 388-412.

[212] Darabi, F. and Wesson, P.S. (2002). Gravitational Conformal Invariance and Coupling Constants in Kaluza-Klein Theory, *Phys. Lett.*, B527, pp. 1-8.

[213] Wesson, P.S. (2002). On Higher-Dimensional Dynamics, *J. Math. Phys.*, 43, pp. 2423-2438.

[214] Fukui, T., Seahra, S.S. and Wesson, P.S. (2002). A Non-Separable Solution of the Vacuum 5D Space-Time-Matter Universe, in Sato, K. and Shiromizu, T. (eds.), *New Trends in Theoretical and Observational Cosmology* (Universal Academic Press, Tokyo), pp. 271-272.

[215] Seahra, S.S. and Wesson, P.S. (2002). The Structure of the Big Bang from Higher-Dimensional Embeddings, *Class. Quant. Grav.*, 19, pp. 1139-1155.

[216] Wesson, P.S. (2002). Classical and Quantized Aspects of Dynamics in Five-Dimensional Relativity, *Class. Quant. Grav.*, 19, pp. 2825-2833.

[217] Wesson, P.S. (2002). The Shape of the Universe, *Astron. Geophys.*, 43 (6), pp. 13-16.

[218] Wesson, P.S. (2002). Five-Dimensional Relativity and Two Times, *Phys. Lett.*, B538, pp. 159-163.

[219] Overduin, J.M. and Wesson, P.S. (2003). *Dark Sky, Dark Matter* (Institute of Physics, London), 216pp.

[220] Wesson, P.S. (2003). *Cosmic Dreams* (Vantage Press, New York), 209 pp.

[221] Wesson, P.S. (2003). Can You Feel It? *New Scientist*, 178 (17 May 2003), pp. 30-31.

[222] Wesson, P.S. (2003). The 4D Klein-Gordon, Dirac and Quantization Equations from 5D Null Paths, *Gen. Rel. Grav.*, 35, pp. 111-119.

[223] Wesson, P.S. (2003). Geophysical Consequences of Modern Cosmologies, in Scalera, G. and Jacob, K.-H. (eds.), *Why expanding Earth?* (Inst. Nazionale Geofisica Vulcanologia, Rome), pp. 411-416.

[224] Seahra, S.S. and Wesson, P.S. (2003). Application of the Campbell-Magaard Theorem to Higher-Dimensional Physics, *Class. Quant. Grav.*, 20, pp. 1321-1339.

[225] Wesson, P.S. (2003). The Equivalence Principle as a Symmetry, *Gen. Rel. Grav.*, 35, pp. 307-317.

[226] Wesson, P.S. (2003). Mass, Particles and Waves in Higher Dimensions, *Int. J. Mod. Phys.*, D12, pp. 1721-1727.

[227] Seahra, S.S. and Wesson, P.S. (2003). Universes Encircling 5-Dimensional Black Holes, *J. Math. Phys.*, 44, pp. 5664-5680.

[228] Liko, T., Overduin, J.M. and Wesson, P.S. (2004). Astrophysical Implications of Higher-Dimensional Gravity, *Sp. Sci. Rev.*, 110, pp. 337-357.

[229] Wesson, P.S. (2004). Space-Time Uncertainty from Higher-Dimensional Determinism, *Gen. Rel. Grav.*, 32, pp. 451-457.

[230] Wesson, P.S. (2004). The Light Stuff, *New Scientist*, 184 (20 Nov. 2004), pp. 31-33.

[231] Mashhoon, B. and Wesson, P.S. (2004). Gauge-Dependent Cosmological 'Constant,' *Class. Quant. Grav.*, 21, pp. 3611-3620.

[232] Wesson, P.S. (2004). Is Mass Quantized? *Mod. Phys. Lett.*, A19, pp. 1995-2000.

[233] Overduin, J.M. and Wesson, P.S. (2004). Dark Matter and Background Light, *Physics Reports*, 402, pp. 267-406.

[234] Liko, T. and Wesson, P.S. (2005). The Big Bang as a Phase Transition, *Int. J. Mod. Phys.* A20, pp. 2037-2045.

[235] Seahra, S.S. and Wesson, P.S. (2005). The Universe as a Five-Dimensional Black Hole, *Gen. Rel. Grav.*, 37, pp. 1339-1347.

[236] Wesson, P.S. (2005). The Spins of Astrophysical Systems and Extended Gravitational Theory, *Astrophys. Sp. Sci.*, 299, pp. 317-327.

[237] Liko, T. and Wesson, P.S. (2005). An Exact Solution of the Five-Dimensional Einstein Equations with Four-Dimensional de Sitter-Like Expansion, *J. Math. Phys.*, 46, p. 062504 (4 pp).

[238] Wesson, P.S. (2005). The Equivalence Principle as a Probe for Higher Dimensions, *Int. J. Mod. Phys.*, D14, pp. 2315-2318.

[239] Wesson, P.S. (2005). Astrophysical Consequences of Extended Cosmology, *Astron. Astrophys.*, 441, pp. 41-45.

[240] Wesson, P.S. (2005). In Defence of Campbell's Theorem as a Frame for New Physics (preprint, arXiv:gr-qc/0507107), 13 pp.

[241] Wesson, P.S. (2006). *Five-Dimensional Physics* (World Scientific, Singapore), 222 pp.

[242] Halpern, P. and Wesson, P.S. (2006). *Brave New Universe* (Joseph Henry Press, Washington), 264 pp.

[243] Wesson, P.S. (2006). Vacuum Instability, *Found. Phys. Lett.*, 19, pp. 285-291.

[244] Ponce de Leon, J. and Wesson, P.S. (2006). An Exact Self-Similar Solution for an Expanding Ball of Radiation, *Int. J. Mod. Phys.*, D15, pp. 395-404.

[245] Wesson, P.S. (2006). Wave Mechanics and General Relativity: A Rapprochement, *Gen. Rel. Grav.*, 38, pp. 937-944.

[246] Wesson, P.S. (2006). Enter the Void, *New Scientist*, 189 (11 Feb. 2006), pp. 32-35.

[247] Zhang, C., Liu, H., Xu, L. and Wesson, P.S. (2006). Universe Evolution in a 5D Ricci-Flat Cosmology, *Mod. Phys. Lett.*, A21, pp. 571-579.

[248] Liu, M., Liu, H., Xu, L. and Wesson, P.S. (2006). Radiation and Potential Barriers of a 5D Black String Solution, *Mod. Phys. Lett.* A21, pp. 2937-2945.

[249] Wesson, P.S. (2007). The Meaning of Dimensions, in Petkov, V. (ed.), *Relativity and the Dimensionality of the World* (Springer, Berlin), pp. 1-11.

[250] Mashhoon, B. and Wesson, P.S. (2007). An Embedding for General Relativity and its Implications for New Physics, *Gen. Rel. Grav.*, 39, pp. 1403-1412.

[251] Overduin, J.M., Wesson, P.S. and Mashhoon, B. (2007). Decaying Dark Energy in Higher-Dimensional Gravity, *Astron. Astrophys.*, 473, pp. 727-731.

[252] Overduin, J.M. and Wesson, P.S. (2008). *The Light/Dark Universe* (World Scientific, Singapore), 225 pp.

[253] Wesson, P.S. and Anderson, M.K. (2008). The Gravity Probe-B Bailout, *Spectrum I.E.E.E.*, 10, pp. 1-4.

[254] Wesson, P.S. (2008). A New Look at the Big Bang, *Int. J. Mod. Phys.*, D17, pp. 635-639.

[255] Wesson, P.S. (2008). The Geometrical Unification of Gravity with its Source, *Gen. Rel. Grav.*, 40, pp. 1353-1365.

[256] Ponce de Leon, J. and Wesson, P.S. (2008). A Class of Anisotropic Five-Dimensional Solutions for the Early Universe, *Europhys. Lett.*, 84, p. 20007, 5 pp.

[257] Wesson, P.S., Mashhoon, B. and Overduin, J.M. (2008). Cosmology with Decaying Dark Energy and Cosmological 'Constant,' *Int. J. Mod. Phys.* D17, pp. 2527-2533.

[258] Wesson, P.S. (2008). Particles versus Waves: Exorcizing a Physics Devil and Peering into the Human Mind (preprint, 5DSTM.org), 12 pp.

[259] Wesson, P.S. (2008). Does God Play Dice with the Universe? (preprint, 5DSTM.org), 9 pp.

[260] Mashhoon, B., Overduin, J.M., Ponce de Leon, J. and Wesson, P.S. (2008). Passing Beyond Gravity (obituary for Hongya Liu, 5DSTM.org), 4 pp.

[261] Wesson, P.S. (2008). Mass and Machian General Relativity (preprint, arXiv:gen-ph/0811.2529), 19 pp.

[262] Wesson, P.S. (2008). Quantization in Spacetime from Null Paths in Higher Dimensions (preprint, arXiv:gen-ph/0812.2254), 15 pp.

[263] Wesson, P.S. (2009). Classical Universes and Quantized Particles from Five-Dimensional Null Paths (preprint, arXiv:gr-qc/0903.2678), 33 pp.

[264] Wesson, P.S. (2009). Consequences of Kaluza-Klein Covariance (preprint, arXiv:gr-qc/0905.0113), 22 pp.

[265] Wesson, P.S. (2010). Time as an Illusion, in Petkov, V. (ed.), *Minkowski Spacetime: A Hundred Years Later* (Springer, Berlin), pp. 307-317.

[266] Wesson, P.S. (2010). Multiple Vacuum States from Higher-Dimensional Relativity: A Short Discussion (preprint, 5DSTM.org), 10 pp.

[267] Wesson, P.S. (2010). Panspermia, Past and Present: Astrophysical and Biophysical Conditions for the Dissemination of Life in Space, *Sp. Sci. Rev.,* 156, pp. 239-252.

[268] Wesson, P.S. (2010). The Scalar Field of 5D Gravity and the Higgs Field of 4D Particle Physics: A Possible Connection (preprint, arXiv:gr-qc/1003.2476), 32 pp.

[269] Wesson, P.S. (2010). Physical Identifications for the Algebraic Quantities of Five-Dimensional Relativity (preprint, arXiv:gr-qc/1006.3529), 33 pp.

[270] Wesson, P.S. (2010). The Embedding of General Relativity in Five-Dimensional Canonical Space: A Short History and a Review of Recent Physical Progress (preprint, arXiv:gr-qc/1011.0214), 20 pp.

[271] Wesson, P.S. (2010). Dynamical Implications of Adjustments to Proper Time Caused by Higher Dimensions: A Note (preprint, arXiv:gen-ph/1011.2791), 13 pp.

[272] Wesson, P.S. (2010). Quantum-Mechanical Waves in Closed Vacuum States (preprint, arXiv:gen-ph/1012.6036), 15 pp.

[273] Wesson, P.S. (2011). *Weaving the Universe* (World Scientific, Singapore), 204 pp.

[274] Wesson, P.S. (2011). *Gambling with Galaxies* (Vantage Press, New York), 209 pp.

[275] Wesson, P.S. (2011). Necropanspermia, *Observatory*, 131 (1221), pp. 63-66.

[276] Wesson, P.S. (2011). "To See the World in a Grain of Sand": A Vintage Idea in Philosophy is Realized in a Modern Theory of Physics, *Observatory*, 131 (1222), pp. 134-138.

[277] Wesson, P.S. (2011). General Relativity and Quantum Mechanics in Five Dimensions, *Phys. Lett.* B701, pp. 379-383.

[278] Wesson, P.S. (2011). The Cosmological 'Constant' and Quantization in Five Dimensions, *Phys. Lett.* B706, pp. 1-5.

[279] Wesson, P.S. (2011). The Physical Nature of Five-Dimensional Solitons: A Survey (preprint, arXiv:gr-qc/1104.3244), 25 pp.

[280] Wesson, P.S. (2011). Particle Masses and the Cosmological 'Constant' in Five Dimensions (preprint, arXiv:gr-qc/1111.4698), 23 pp.

[281] Wesson, P.S. (2011). *Tales of Crazy Scientists* (unpublished collection of 12 science-fiction short stories, available from interstellarundertakers.com/short_stories.htm)

[282] Mashhoon, B. and Wesson, P.S. (2012). Mach's Principle and Higher-Dimensional Dynamics, *Ann. Phys.* (Berlin), 524, pp. 63-70.

[283] Mashhoon, B. and Wesson, P.S. (2012). Mach, the Universe and Foundations of Dynamics, *Ann. Phys.* (Berlin), 524, p. A44.

[284] Wesson, P.S. (2012). Constructing Reality: Quantum Theory and Particle Physics (book review). *Observatory*, 132 (1226), pp. 46-47.

[285] Wesson, P.S. (2012). Astronomy and the Fifth Dimension, *Observatory*, 132 (1231), pp. 372-376.

[286] Wesson, P.S. (2012). Particles, Waves and Vacuum in Five Dimensions: A Status Report (preprint, arXiv:gr-qc/1205.4452), 22 pp.

[287] Wesson, P.S. and Overduin, J.M. (2012). A Scalar Field and the Einstein Vacuum in Modern Kaluza-Klein Theory (preprint, arXiv:gr-qc/1209.5703), 13 pp.

[288] Wesson, P.S. (2013). Vacuum Waves, *Phys. Lett.* B722, pp. 1-4.

[289] Wesson, P.S. (2013). Transformity: The Dependence of the Laws of Physics on Higher-Dimensional Coordinate Transformations, *Am. J. Sp. Sci.* 1 (2), pp. 54-60.

[290] Wesson, P.S. and Overduin, J.M. (2013). Scaling Relations for the Cosmological 'Constant' in Five-Dimensional Relativity, *Adv. High Energy Phys.*, 2013, p. 214172, 6 pp.

[291] Overduin, J.M., Everett, R.D. and Wesson, P.S. (2013). Constraints on Kaluza-Klein Gravity from Gravity Probe B, *Gen. Rel. Grav.*, 45, pp. 1723-1731.

[292] Wesson, P.S. and Overduin, J.M. (2013). Wave Mechanics and the Fifth Dimension (preprint, arXiv:gr-qc/1302.1190), 14 pp.

[293] Wesson, P.S. and Overduin, J.M. (2013). The Scalar Field Source in Kaluza-Klein Theory (preprint, arXiv:gr-qc/1307.4828), 7 pp.

[294] Wesson, P.S. (2014). Einstein's Equations, Cosmology and Astrophysics, in Ashtekar, A. and Petkov, V. (eds.), *The Springer Handbook of Spacetime* (Springer, Dordrecht), pp. 617-627.

[295] Wesson, P.S. (2014). Looking for the Fifth Dimension, *Physics International,* 5 (1), pp. 5-7.

[296] Wesson, P.S. (2014). The Dispersion Relation for Matter Waves in a Two-Phase Vacuum, *Mod. Phys. Lett.* A29, pp. 1450168-76.

[297] Wesson, P.S. (2014). Higher-Dimensional Communication and S.E.T.I. (preprint, arXiv:gen-ph/1401.2883), 8 pp.

[298] Wesson, P.S. (2014). Foreword for Report on the Relativity Theory of Gravitation by A.S. Eddington (reprinted by the Minkowski Press, Montreal), pp. iii-iv.

[299] Wesson, P.S. (2015). Quantum Uncertainty and the Fifth Dimension, *Physics International,* 2015, 6 (2): 78-81.

[300] Wesson, P.S. (2015). The Status of Modern Five-Dimensional Gravity (A Short Review: Why Physics Needs the Fifth Dimension), *Int. J. Mod. Phys.* D24, p. 1830001, 11 pp.

[301] Wesson, P.S. (2015). Where is the Fifth Dimension? (preprint, 5DSTM.org), 10 pp.

[302] Wesson, P.S. (2015). An Alternative Approach to Quantum Mechanics Based on Five Dimensions (preprint, 5DSTM.org), 19 pp.

[303] Wesson, P.S. and Overduin, J.M. (2015). Waves and Causality in Higher Dimensions, *Phys. Lett.*, B750, pp. 302-305.

[304] Wesson, P.S. (2015). A Machian Definition of Particle Mass in Higher-Dimensional Gravity (preprint, arXiv:gr-qc/1602.04390), 16 pp. (posthumous).

[305] Wesson, P.S. (2016). Extra Dimensions and Phase Experiments, *Annalen der Physik* (Berlin), 528, pp. 307-312 (posthumous).

Index